W9-CRJ-293

Praise for

PANDORA'S LUNCHBOX

"A gripping exposé."

—*The Wall Street Journal*

"Indispensable."

—*Newsday*

"Fascinating."

—The A.V. Club

"So much fun that you might forget how depressing it all is. . . . There are more *Holy Cow!* moments here than even someone who thinks he or she knows what's going on in food production could predict."

—Mark Bittman, *The New York Times*

"This absolutely fascinating—and rather infuriating—look at what society is really eating is a must for any responsible adult."

—*Publishers Weekly*

"Warner puts forth compelling evidence that proves we, as a society, place a premium on convenience over health when it comes to food. And the effects of that prioritization—whether intentional or not—can be pretty disastrous."

—*Huffington Post*

"*Pandora's Lunchbox* is the sort of electrifying journalism that comes around only once in a while. Read it. Especially if you're experiencing fighting-the-good-fight fatigue and starting to doubt that buying a head of lettuce at the farmers' market is in fact a radical act—a health-giving, world-fixing, dead-zone-opposing radical act in a world where our daily 'bread' can involve the manufacture of foamy plastics."

—Experiencelife.com

"In *Pandora's Lunchbox*, Melanie Warner has produced an engaging account of how today's 'food processing industrial complex' replaced real foods with the inventions of food science. Her history of how this happened and who benefits from these inventions should be enough to inspire everyone to get back into the kitchen and start cooking."

—Marion Nestle, Professor of Nutrition,
Food Studies, and Public Health at New York University and
coauthor of *Why Calories Count: From Science to Politics*

"*Pandora's Lunchbox* is a brilliant and fascinating exploration of how our food gets processed, its powerful effects on our health, and what we can do about it. Highly recommended!"

—Dean Ornish, MD, author of *Eat More, Weigh Less*
and *The Spectrum: A Scientifically Proven Program to Feel Better,
Live Longer, Lose Weight, and Gain Health*

"Melanie Warner is a journalist of keen skill, and in *Pandora's Lunchbox* she pries the lid off well-packaged secrets about how our so-called food is made. The resulting bounty of insights and revelations is almost overwhelming. This is a book of stunning, at times shocking truths, told in a crisp, compelling narrative. Of profound importance for everyone who eats."

—David L. Katz, MD, MPH, FACPM, FACP,
Director of Yale University Prevention Research Center and
Director of Integrative Medicine Center at Griffin Hospital

"Warner pulls back the curtain to reveal the industry secrets of how our most basic staples are being transformed into processed foodstuffs to boost profits. We get an (un)healthy dose of hexane-extraction, gun puffing, and roast chicken–type flavor, but like the best investigative journalists, she uses the personal stories of food scientists, innovators, and crusaders—not to mention her own home experiments—to show why you'll want to think twice before hitting the drive-thru or reaching for that 'health bar.'"

—Robert Kenner, director of *Food, Inc.*

PANDORA'S LUNCHBOX

HOW PROCESSED FOOD TOOK OVER THE AMERICAN MEAL

Melanie Warner

SCRIBNER

New York London Toronto Sydney New Delhi

Scribner
A Division of Simon & Schuster, Inc.
1230 Avenue of the Americas
New York, NY 10020

Copyright © 2013 by Melanie Warner

All rights reserved, including the right to reproduce this book or
portions thereof in any form whatsoever. For information, address
Scribner Subsidiary Rights Department, 1230 Avenue of
the Americas, New York, NY 10020.

First Scribner trade paperback edition February 2014

SCRIBNER and design are registered trademarks of The Gale Group, Inc.,
used under license by Simon & Schuster, Inc., the publisher of this work.

For information about special discounts for bulk purchases,
please contact Simon & Schuster Special Sales at 1-866-506-1949 or
business@simonandschuster.com.

The Simon & Schuster Speakers Bureau can bring authors to your
live event. For more information or to book an event, contact the
Simon & Schuster Speakers Bureau at 1-866-248-3049 or
visit our website at www.simonspeakers.com.

Book design by Ellen R. Sasahara

Manufactured in the United States of America

1 3 5 7 9 10 8 6 4 2

Library of Congress Control Number: 2012042352

ISBN 978-1-4516-6673-1
ISBN 978-1-4516-6674-8 (pbk)
ISBN 978-1-4516-6675-5 (ebook)

For Therese and Rich

I have always stood for food that is food.

—Harvey Wiley, pioneering chemist
and father of the Food and Drug
Administration

Table of Contents

Introduction

A NUMBER OF YEARS AGO, I went to the supermarket and bought an overflowing armful of cereal boxes and cookie packages. I'd started writing about the food industry for the *New York Times* not long before, and I'd decided to test whether those expiration dates printed on packages actually meant anything. I'd always wondered what happened to food after the expiration dates passed. Would the cookies turn green or taste like old shoes? Would bugs crawl out of the cereal? I tucked the boxes and crinkly bags away in my kitchen for nearly a year. The dates printed on the packages came and went, and when I opened them, the results were fairly unremarkable: my cereal and cookies looked and tasted perfectly normal, almost as if I'd just bought them.

I started wondering how long other foods would last. My experiment expanded—frozen dinners, kids' lunches, loaves of bread, processed cheese, hot dogs, pudding, and Pop-Tarts. I brought home samples of fast-food burgers, fries, chicken sandwiches, and chicken nuggets. At the time I was working from home, and I had to keep everything out of reach of our two young sons, who were never able to understand why they couldn't eat just one Oreo or have a taste of a Pop-Tart.

I worried that my work area might succumb to some sort of awful infestation. I pictured fruit flies or those tiny worms that get into the forgotten bag of flour in the back corner of the top cabinet. But none of this happened. Much of my collected food stubbornly refused to decay, even after as many as six years—far beyond expiration dates.

I wondered what had happened to this food to make it so eternal, so unappealing to the mold and bacteria that normally feast on ignored leftovers and baked goods. It seemed to me that the dates printed on the package had little to do with true "expiration." What did those dates actually mean? How was it possible that foods that seemed perfectly edible could be immune to natural processes of decomposition? What were we actually feeding our kids?

Around our house, my experiments were regarded as little more than mildly amusing, sort of weird, and definitely gross. My food collection was a funny little hobby. Until the guacamole incident.

On a Fourth of July trip out of town in 2011, my husband had returned from the grocery store with a tub of "fresh guacamole." "They made an announcement over the loudspeaker that they had just made it over at the deli, so I went and got some," he said proudly.

The container had a haphazardly applied sticker on it, indicating that it very well could have been "made fresh" by one of the store's white-coated deli workers. But there was something unusual about the ingredients: Hass avocados, salt, ascorbic acid, citric acid, xanthan gum, amigum, text-instant, tomatoes, yellow onion, jalapeño, cilantro.

I was knee-deep in research on food additives, but I'd never heard of amigum or text-instant. I went to the store and bought another tub, tucking it into our fridge at home and figuring I'd look into those strange ingredients later. Mostly I forgot about it. Then, nine months later, my mom, who lives with us in Boulder, Colorado, announced she'd tried some of the guacamole. We'd just had a birthday party for one of our boys, and I'd bought some dips from Whole Foods. I hoped that was what she was referring to, but I was pretty sure all of it was gone.

My mom had tried the other guacamole, the Fourth of July stuff, of course. "It was a little spicy," she declared.

My food museum was nauseating, but it had never occurred to me that it could actually sicken anyone. I was concerned because, as an older person, my mom has a higher risk of contracting a life-threatening food-borne illness. Mom assured me everything would be fine; she is nothing if not an unrelenting optimist. Amazingly, though, she was right. Not even an intestinal rumbling. She'd only tried a little, thank God.

Some people probably would have looked at that tub of green goop and not eaten any of it. It was brown around the edges and didn't look particularly fresh. But others might have done exactly what my mom did, and mistaken it for something edible. Even homemade guacamole tends to darken after a few days, and what my mom ate had none of the red flags that help guide us in our decisions about whether or not to consume something. There was no mold and no bad smell.

Like so much of the food we eat today, this immortal guacamole was not what it seemed. It had, in fact, been prepared—or assembled—by those deli workers, but not according to any recipe you'd use at home. It didn't look like a processed food, but that's exactly what it was. Along with the usual avocados, tomatoes, and onions, this guacamole had corn. Or corn manipulated beyond recognition so that it had been transformed into preservatives you can't taste, smell, or see. And then there was that "text-instant," as well as "amigum"—an ingredient that, I later learned, was even more bizarre than I could have imagined.

And that is the story of so much of our food, it turns out. Although my mother instilled in me a healthy skepticism of processed foods growing up, allowing me very limited access to what she called "gooped-up" food, I had no idea just how tremendously technical our food production had become until my food experiments impelled me to take a closer look. What started as an earnest attempt to understand the true meaning of labeling on the packages of the foods so many of us eat became a larger journey that brought me inside the curious, intricate world of food science and technology, a place where food isn't so much cooked as disassembled and reassembled. Over the last century, such complex modes of production have ushered in a new type of eating, what we call processed food.

Considering our vast and bewildering cornucopia of modern food choices, it's easy to forget that most of the items lining the inner aisles of the supermarket and the substances offered on fast-food menu boards simply didn't exist a century ago. The avalanche of prefabbed, precooked, often portable food into every corner of American society represents the most dramatic nutritional shift in human history. If we really are what we eat, then Americans are a different dietary species from what we were at the turn of the twentieth century. As a population, we ingest double the amount of added fats, half the fiber, 60 percent more added sugars, three and a half times more sodium, and infinitely greater quantities of corn and soybean ingredients than we did in 1909.

The trouble with this wholesale remaking of the American meal is that our human biology is ill equipped to handle it. The way our bodies metabolize food is stuck somewhere in the Stone Age, long before the age of Cheez Whiz, Frosted Flakes, and Classic Chick'N Crisp fried in vegetable oil. Our many novel and high-tech manipulations of food destroy much of its essential geography, resulting in all sorts of unintended consequences. When we start taking food apart and industrially processing it, it often stops making biological sense.

Processed food is even more ubiquitous than we think it is, in part because many products are designed to look as if they're not really processed at all. Subway's "fresh" sandwiches and the center aisles at Whole Foods, for instance, can both be quite perplexing. What are boxes of General Mills's Cascadian Farm's Fruitful O's and Cinnamon Crunch, if not Froot Loops and Cinnamon Toast Crunch by other names? Whole Foods co-founder John Mackey once acknowledged that some of what his stores sell is a "bunch of junk." And Subway's bread is not much more fresh and its meat no more whole than the bags of chips sitting up at the register. In total, some 70 percent of our calories come from this sort of (ultra) processed food. As an industry, this amounts to $850 million a year.

And yet many foods that some might call processed in fact are not. At one point during my research, I attended an industry conference where the keynote discussion sought to tackle the merits of food processing. The example most often cited was pasteurized milk. Thank goodness for food scientists, the argument went, who save Americans from countless outbreaks of campylobacter and *E. coli*. Yes, thank goodness, but pasteurized milk, let's be clear, is not a processed food. Nor are frozen peas, canned beans, washed and boxed spinach, bags of baby carrots, packages of aged cheese, or boxes of raw, frozen ground beef shaped into hamburgers.

At one point in time these products undoubtedly would have been heralded as newfangled creations. But today they barely register on the processing continuum and are not included in that 70 percent figure, which comes from a rigorous analysis done by the Brazilian nutrition scientist Carlos Monteiro. As a general rule—in a universe of tens of thousands of foods, there are always exceptions—a *processed food* is something that could not be made, with the same ingredients, in a home kitchen. Your home kitchen.

I've written this book with the core belief that it's important to understand what we're eating. Some people won't want to know and would rather just keep eating all their favorite foods in peace, and this book isn't for them. But for those who believe in the virtues of a health-promoting diet for themselves and their families, few things are more important to understand than what happens to our food before it gets to our plates—whether it's arrived from the farm reasonably intact or has had a long, multibranched journey through the nutritionally devastating food-processing industrial complex.

The aging guacamole notwithstanding, my mom, who read food labels with a discriminating eye long before it was fashionable, still does her best to avoid "gooped-up" food. She cooks most of what she eats and continues to survey ingredients (although apparently not if the food is already in the fridge). But her diet isn't one of deprivation. She eats meat and dairy and plenty of butter. She's never been lactose-free, sugar-free, caffeine-free, or fat-free. Nor does she have any plans to go gluten-free: the woman eats more bread than anyone. The only organizing principle of her diet is that she predominantly consumes things she would have recognized as food growing up in the thirties in Nova Scotia. She doesn't eat fast food; there was none back then. And she's never owned a microwave; they weren't available until the seventies.

It seems to have worked well for her. In her early eighties, she's in near-perfect health, with no chronic conditions and no prescriptions to fill, something that, if you ask her, she will attribute in no small part to what she eats. "What you put into your body matters, Melanie," she told me more than once while I was in college, eating Pop-Tarts and pizza for dinner. "Just because it's edible doesn't mean it's good for you."

As hard as it was to acknowledge at the time, she was on to something.

1

Weird Science

Eight percent of U.S. kids have food allergies. Luckily very little of what they eat is technically food.

—Stephen Colbert

On a swampy day in New Orleans, 15,000 people streamed into the Morial Convention Center, an immense structure on the banks of the Mississippi. Food scientists, chemists, research and development chiefs, marketing executives, salespeople, professors, and students, they wore badges attached to ribbons strung around their necks and hauled around thick packets of information in free shoulder bags festooned with company logos. Old acquaintances greeted one another with backslaps and squeals of delight. For three days in mid-June, the convention center, just upriver from the elegant plantation homes and leafy throughways of the Garden District, hummed with a low roar of chatter.

An annual event since 1940, the Institute of Food Technologists' yearly meeting, known as IFT, is the country's largest and most anticipated gathering in the processed food industry. There are few locations better suited to a celebration of food than New Orleans. The town's luscious aromas are an indelible a part of its character, with New Orleans's contributions to the national diet including crayfish bisque, jambalaya, po' boys, spicy red beans and rice, oysters Rockefeller, shrimp remoulade, trout amandine, and shrimp cooked every which way. These carnal, spicy foods have evolved from a rich brew of diverse populations and prompted the novelist Tom Robbins to write, "The minute you land in New Orleans, something wet and dark leaps on you and starts humping you like a swamp dog in heat, and the only way to get that aspect of New Orleans off you is to eat it off."

There was lots to eat at IFT, but not the standard New Orleans fare. IFT isn't the Fancy Food Show. You'll find no morel mushrooms, Italian fruit vinegar, or Hawaiian honey. And it isn't Aspen's Food & Wine Classic with its melted-cheese master class. Dedicated to the latest advances in food science, it's a conference where the idea of eating seemed to take on a curious notional quality, a place where food isn't plant or animal but a *matrix* or *application*.

The conference opened with a video entitled "Day in the Life of a Food Scientist," featuring both a NASA food scientist talking about how she confects space meals and scientists at Disney making new kids' snacks with cartoon characters. Michael Specter, author of the book *Denialism: How Irrational Thinking Hinders Scientific Progress, Harms the Planet, and Threatens Our Lives,* delivered the keynote address. He spoke about how science—food and otherwise—is often misunderstood by the public. Afterward, a panel discussion explored "Changing the Image of Food Science in the Marketplace."

The conference's main event was held in an immense exposition hall, where nine hundred companies that supply ingredients for processed food had set up booths to showcase their new products. The largest installations featured colorful banners suspended from the ceiling, plush carpets, comfy couches, counters with bar stools, and small kitchens. One company had driven a truck into the hall to serve as its cooking station. Another hung a giant salt-shaker from the rafters at an angle, as if it were about to unleash a snowstorm on the crowd below. Along the perimeter, rows of more modest setups extended as far as the eye could see.

As I wandered the floor, I noted strange banners that advertised "cheese application needs" and "emulsified meat systems." A company selling milk powders declared, "At Marron Foods, people eat, sleep, and drink agglomeration." Another firm boasted about "meat enhancement." At one booth, I asked a twenty-something sales rep for a dairy ingredient company whether it was difficult to explain to his friends what he does for a living. "I tell them that I'm building milk backwards," he said, grinning.

It was, I thought, an apt description of the basic blueprint for processed food. The companies in the convention center disassemble food (usually corn, soybeans, wheat, or milk) into hundreds of different ingredients, which manufacturers such as Kraft, Pepsi, General Mills, ConAgra, Tyson, and Sysco then construct into the packaged foods we buy at grocery stores and fast-food restaurants. Some of the ingredients on display were ordinary and familiar—I saw a company selling vanilla extract and another displaying dried fruit. Many others were novel creations designed to perform highly specialized functions: monk fruit extract to replace sugar, specialized yeast extracts to lower salt, algae-based flour to reduce fat.

Many ingredients I'd never envisioned myself eating, but probably have—inner pea fiber, microparticulated whey protein concentrate, corn fiber designed to be dissolved into clear beverages. I saw a company selling a substance made from castor oil and added to chocolate to lower costs. A Chinese firm was offering promotional samples of synthetic fruit flavoring. And there was plenty of that old standby, xanthan gum, the slimy coating produced by fermentation of the bacteria *Xanthomonas campestris* with corn syrup.

Although IFT is a conference about processed food, no one who goes there refers to it that way. For those who work in the industry, the term is vague and prosaic, if not pejorative. The term doesn't capture the complexity and breadth of their business. To articulate accurately the sophistication of manufactured food— whether a frozen dinner, package of lunch meat, cereal bar, or Egg McMuffin—a much more precise and technical language is preferred. This opaque vernacular was on display on the convention center's upper level, where a three-day lineup of meetings and panels was under way. Some were accessible to nonscientists, like one entitled "Reducing Sodium in Foods: Implications for Flavor and Health." Others required substantial translation. A dairy-food scientist from a company called TIC Gums gave a talk called "Texturing Alternatives for All-Natural Dairy Products Using Synergistic Hydrocolloids," and a rep from a scientific standards organization held forth on "Developing a Compendial HPLC Procedure for Steviol Glycosides."

Even with superhuman levels of energy, visiting even a fraction of the small city of booths and exhibits would require more than the three full days. At every turn, a smiling person offered a tray of mouthwatering samples or motioned toward a counter lined with treats. Having eaten my way around the floor, I remarked to one of the reps in the IFT pressroom that I was going to cut myself off. "Are you feeling okay?" she asked, a bit worried. I wasn't sure what she meant. "Some of these things are really new and they don't always agree with people. I had some problems last year."

Cheaper Ingredients

I tasted a blueberry muffin baked with something called Flav-R-Bites that were moist and sweet and tasted a lot like blueberries. As I chewed, a sales rep from Cereal Ingredients split open a muffin to show me what his product looked like inside. "Like a lot of food, it's about eye appeal. You want it to look just like blueberries."

Actual berries are quite expensive. Flav-R-Bites consist of flour, sugar, starch, flavorings, and just six percent blueberry solids, but enough so that the word "blueberry" can appear on the label. Such substitutions help keep raw material costs low and ensure an end-less and affordable national supply of blueberry muffins, scones, and bagels. The nuggets also have a much longer shelf life; with bona fide blueberries, you get maybe a few days of longevity. Cereal Ingredients had dozens of nuggets on display in every imaginable flavor and color. Lined up in glass jars, they shimmered.

Over at the other end of the expo hall, I wandered into National Starch's installation, one of the show's biggest and most promi-nent. Founded back in 1890 in Bridgewater, New Jersey, the com-pany provides food starches to customers such as General Mills, Nestlé, Kellogg's, and McDonald's. National Starch's products first ushered in the era of frozen meals in the fifties. Without modified starches, the sauces in TV dinners would have been a goopy, oily mess and the meat dry and rubbery. Today, the com-pany's starches, which are made from corn, tapioca, and potatoes, still add structure to sauces and moisture to meat. They also give yogurts and puddings the sort of thickness you can plant a spoon into, provide frozen food with "freeze-thaw stability," and, perhaps most important, help lower production costs for food products of all stripes.

These altered starches do this very well, apparently. According to an ambitious campaign National Starch ran a number of years ago, its starches, which were branded with the trademarked name "Starchology," can mimic a variety of more traditional food ingredients. "The tomato is feeling insecure," read one trade magazine ad. "Starchology can squeeze 40% out of vegetable solids. It's tough for the tomato but terrific for you." Another ad targeted manufacturers laboring under the unnecessary burden of buying real butter: "Butter is feeling left out. Starchology can help you replace fat with savings." As an added bonus, these food-simulating starches could be identified on package labels by the everyday, reassuring words "cornstarch" or "flour." There was no need to indicate that these starches had been altered in labs, either by chemicals or through a heating and cooling process. The company explained that "Wholesome, consumer-friendly ingredients enhance your products and give them a 'made-at-home' feel while withstanding typical food processing conditions."

You might think that having a product contain actual tomatoes or real blueberries would be a good thing. But when processed food is concerned, fruits and vegetables cause problems since they contain water, which can cause spoilage or ice crystals when products are frozen—not to mention that these whole-food ingredients are expensive for food manufacturers. All businesses must be mindful of how operating costs affect the bottom line, and food companies may be under a greater burden than most, since American grocery shoppers and fast-food eaters have become deeply attached to the idea of inexpensive food. The amount we pay for our food has declined dramatically over the last six decades, from 20.6 percent of disposable income in 1950 to now 9.8 percent.

This is lower than at any other time in our history and less than any other country. Most food companies dread the idea of raising prices, since it's certain to be followed by some degree of customer defection.

For IFT 2011, National Starch—which merged in 2012 with a starch and high-fructose corn syrup maker named Corn Products and renamed itself Ingredion—had found an ideal food with which to exhibit the cost-cutting benefits of its starch technology: Greek yogurt. Sales of this yogurt, which is thicker and higher in protein than standard varieties, had catapulted to a quarter of all yogurt sales in just four years, taking big manufacturers like Dannon and General Mills, which owns the Yoplait brand, by complete surprise. Everyone was looking to take advantage of this booming market, yet cost was an issue. To develop its characteristic thickness, Greek yogurt must be strained in $10-million machines—one of the reasons that containers of Greek yogurt can cost twice as much as regular yogurt.

To solve this problem, National Starch devised a prototype Greek yogurt that could be manufactured at a fraction of the cost. The cheaper solution, served in clear plastic cups and lined neatly along a counter in the convention hall, was "Greek-style" yogurt made with its Novation Indulge 3340 tapioca starch and milk protein concentrate supplied by another company. The yogurt was topped with berries, or "superfruits," as the industry has taken to calling them, and it was thick and creamy, with a slightly pasty texture. According to my taste buds, it tasted exactly like Greek yogurt. Paul Petersen, National Starch's tall, slim, New Zealand-born global marketing director for texture products, wandered over to ask if I had any questions.

"How's your starch being used in the yogurt? As a thickener?" I asked.

"It's a texturing system," he said. "We don't like to use the term 'thickeners,' since that implies tough and clumpy. It binds with moisture to give that creamy texture people eating Greek yogurt expect."

"But would those people feel shortchanged if they knew they weren't eating real Greek yogurt?" I asked.

Petersen looked at me as if I'd missed the point entirely. "There's no standard or rule of identity for Greek yogurt, so there is no real thing," he said.

He was right. The Food and Drug Administration maintains regulations for what can go into roughly two hundred eighty different foods—rules that don't include newer products like Greek yogurt. There's no standard for Greek yogurt any more than there are regulations for what constitutes a Greek salad. "And don't forget," Petersen said, "this yogurt is going to cost much less than the traditional Greek yogurts."

In fact, it already did. "Greek yogurts" containing National Starch's thickeners and added milk protein concentrate were on the market, and they cost less. Safeway's store brand Lucerne had one, as did Yoplait. However, several months after IFT, a General Mills food scientist told me that this was one cost-cutting move that hadn't quite turned out as hoped. Yoplait Greek yogurt wasn't selling particularly well because customers perceived it to be less authentic than other brands. "The Greek people who work here think it's terrible," she added. The company was considering going back to the drawing board to do actual straining, she said. As of the fall of 2012, General Mills seemed to be still testing the benefits of this new authenticity. Some packages of Yoplait Greek yogurt were made with milk protein concentrate; others weren't.

After finishing my "Greek yogurt," I moved on to National Starch's light cucumber ranch dip, made with their Precisa Cling

20 starch. Petersen explained what was happening here: "You want the dip to cling to the vegetables but not to have the consistency of snot." I tried some. The dip formed a nice, tight ball around my baby carrot, nothing close to a drippy homemade veggie dip situation. I wiggled it a little and, as advertised, the dip clung.

White Powders

I suppose I never realized it would be someone's job to negotiate the fine line between sturdy and snotty veggie dips. Or to measure moisture and fat "cook-out" in hamburgers, as a rep at International Fiber Corporation (IFC) put it. "The demo is amazing," he told me as an enormous poster of a thick hamburger dangled over his head. The company's oat fiber, he explained, helps prevent burgers from shrinking when they're cooked, making them juicier and allowing less meat to be used for that quarter pounder. One of IFC's ingredients—at the time *isolated oat product*, now *cellulose* (made from tree pulp)—is among the purported nonbeef substances in Taco Bell's taco meat that prompted a much-publicized 2011 lawsuit. The suit, which accused Taco Bell of using less than the required level of meat in its tacos, was withdrawn just two months after it was filed. But not before Taco Bell's CEO Greg Creed had a chance to go on *Good Morning America* to talk to George Stephanopoulos, who asked him, "What's an isolated oat product?" Creed was forced to admit he had no idea.

The people who go to IFT every year don't consider any of this as odd. Not even a little. It's their job to sell isolated oat substances and cheaper yogurt ingredients. It's how they pay their mortgages and clothe their families. And beyond that, a lot of people in the food industry quite like their jobs; formulating a new snack bar or frozen dessert can yield the same pleasures as solving a really challenging puzzle. When I asked one food scientist whether he thought the average person would find IFT perplexing, he replied, "Not everyone can eat fresh vegetables." The world needs processed food, he argued—though he admitted he needed it less than most. He told me he likes to shop at farmer's markets and plants a garden every spring.

Perhaps my most surreal moment at IFT came at a large, circular encampment erected by Tate & Lyle, a $4.3 billion British agribusiness that got its start selling sugar in the 1800s. The company now makes a range of ingredients, including starches and various sugar substitutes, such as Splenda. I walked onto a thick, gray carpet and bellied up to a makeshift bar to sample a small dish of vanilla parfait topped with a single raspberry. A sales rep wearing a seafoam-green shirt emblazoned with a Tate & Lyle logo explained that this parfait was already being sold at some supermarkets as a dip for fruit. She said it was made with two different types of Tate & Lyle's corn-based starches and sweetened with its crystalline fructose, also made from corn. "The starches work as a kind of glue that binds everything together," she explained.

Under a sign that read "Our ingredients, your success," I tasted the contents of my parfait cup. It was smooth and sweet, but oddly bland and indistinct. I was at a loss to figure out what exactly I was eating. "What's in it?" I asked. "You know, the primary ingredient."

The rep looked at me with a puzzled, blank stare. She turned to her colleague, who also had no idea. After a few moments, she said, "It's a cultured dairy product."

"So it's yogurt?"

"Um, it's not yogurt." She paused. "It's a powdered product probably. You'd add water to it. But it's definitely cultured dairy. That's where you're getting the tangy flavor."

I wasn't getting much of a tangy flavor, but that was beside the point. The parfait wasn't food so much as the chosen delivery system for several edible powdered ingredients, which, I was coming to realize, were everywhere.

If you strip away the food freebies and colorful backdrops of plump fruits and juicy burgers, IFT stands as a grand festival of neutral-hued powders. Crystalline fructose and modified starches are white, as are Splenda and monk fruit extract. Yeast extracts, enzymes, preservatives such as BHT and citric acid, dough conditioners like ammonium sulfate and sodium stearoyl lactylate, and many flavorings are sold as beige powders. Soy protein is a pale yellow powder, and dairy proteins are closer to white. A lot of synthetic vitamins are white powders—and, of course, that yogurt-esque ingredient in my parfait. The food industry relies heavily on these dried, finely pulverized materials because they're cheap and convenient to ship, and because they last much longer than anything with moisture in it.

You probably don't think of your lunch as being constructed from powders, but consider the ingredients of a Subway Sweet Onion Chicken Teriyaki sandwich. Of the 105 ingredients, 55 are dry, dusty substances that were added to the sandwich for a whole variety of reasons. The chicken contains thirteen: potassium chloride, maltodextrin, autolyzed yeast extract, gum Arabic, salt, disodium inosinate, disodium guanylate, fructose, dextrose, thiamine hydrochloride, soy protein concentrate, modified potato starch, sodium phosphates. The teriyaki glaze has twelve: sodium benzoate, modified food starch, salt, sugar, acetic acid, maltodextrin, corn starch, spice, wheat, natural flavoring, garlic powder, yeast extract. In the fat-free sweet onion sauce, you get another eight: sugar, corn starch, modified food starch, spices, salt, sodium benzoate, potassium sorbate and calcium disodium EDTA. And finally, the Italian white bread has twenty-two: wheat flour, niacin, iron, thiamine mononitrate, riboflavin, folic acid, sugar, yeast, wheat gluten, calcium carbonate, vitamin D2, salt, ammonium sulfate, calcium sulfate, ascorbic acid, azodicarbonamide, potassium iodate, amylase, wheat protein isolate, sodium stearoyl lactylate, yeast extract and natural flavor.

If you were to make this sandwich at home with a basic chicken breast and fresh bread made with minimal ingredients, it would contain only a handful of these things. Mass-scale food processing, however, requires an entirely different system of assembly, one fraught with often conflicting expectations. Manufactured food needs not only to taste good, for instance, but also to withstand the wear and tear of processing. It has to look and taste exactly the same every time. It also has to have a long shelf life, be produced cheaply and efficiently, and on top of all that, it would be nice if it could be marketed as healthy. Modified food starches and sugars are what allow Subway to boast that its Sweet Onion Chicken Teriyaki sandwich has only 4.5 grams of fat.

Such prerequisites present food manufacturers and ingredient companies with no shortage of brain-bending problems, which is one of the reasons the people who decide to become food scientists find their work so fascinating.

Food Science U

Steve Smith is strolling the halls of Purdue University's expansive $28 million Philip E. Nelson food science building, and each time someone approaches, he flashes a wide, gap-toothed grin and pitches his steaks. "Salisbury steaks," he says, "We've got Salisbury steaks downstairs." As head of the food science department's sensory testing lab, he does this sort of thing all the time. Every day except Fridays, Smith oversees two taste tests, each of which requires at least one hundred people to sit at a basement computer and answer questions about two similar but different food items that emerge from a sliding panel in the wall. A heavyset guy with salt-and-pepper hair and a matching mustache, Smith has the sort of unflappable enthusiasm and easy likability that make this job look effortless.

At least, he does most days. Today he's having trouble rounding up the last few volunteers for the Salisbury steaks, which isn't all that surprising. When was the last time you went out to a restaurant and ordered Salisbury steak? The hunks of meat Steve Smith is hawking are made by a Cincinnati-based foodservice company called AdvancePierre and sold to school and hospital cafeterias. An earlier tasting from the same company featured a sodium-reduced burger and had many more takers, including me. Many of the taste tests sell themselves. "People love it when we have cookies or french fries or smoothies," Smith says. "Next week, we have Mrs. Fields's carrot cake and donuts. I'm not expecting any problems with those."

I'd come to Purdue just several weeks after fall classes had started up to understand how one gets into the business of being a food scientist. Located in West Lafayette, Indiana, Purdue is one of only thirty-eight universities around the country that offer undergraduate and graduate degrees in the field. Although there are no rankings for food science the way there are for medicine, law, and business, a tally of leading schools would include Purdue in the top five. The school has an exceptionally productive and collaborative relationship with the food industry. Large manufacturers and ingredient makers have helped create labs and contribute to student scholarships. Their employees appear regularly on the leafy, picturesque campus to give talks to students and interview them for jobs and internships, a process that was already in full swing by mid-September.

During my visit, the university hosted its Industrial Round-table, an annual career fair held on the campus's central outdoor quad. There I met Purdue graduates who are now role models for current students. Rodney Green, a Purdue PhD, works at ConAgra on Hunt's canned tomato products and Van Camp's baked beans. His colleague Kirsten Fletter devises new varieties of Slim Jim, Banquet frozen food and Orville Redenbacher popcorn. Jessica Schroeder got a masters degree in 2007 and is now a food scientist for Pepsi's Tropicana beverages, developing the lower-calorie line of Trop50 juices. "It's got stevia, the natural sweetener, in it," she said, proudly hoisting a bottle from a table lined with PepsiCo products. Lynn Choi Perrin, a senior scientist at General Mills, oversees Pillsbury Toaster Strudels, Totino's individual frozen pizza, and Fruit Roll-Ups. Also on the quad were reps from Kraft, Nestlé, french fry maker McCain Foods, and the flavor company Sensient Technologies.

Even though millions of people have consumed the fruits—or fruitlike substances—of their labor, the careers of people like Perrin and Schroeder are unfamiliar to most. Parents dream of their kids one day becoming doctors, lawyers, software moguls, famous athletes, and—until recently, perhaps—Wall Street investment bankers. Nobody imagines their children as food scientists, unless maybe they're food scientists themselves. "A lot of people think it has something to do with cooking," one scientist lamented. Purdue professor Lisa Mauer told me a story about her well-meaning mother-in-law once giving her a cookbook in which to store class notes, a cute but ironic gesture, since much of the field's research ultimately affords us the convenience of never needing to cook.

Like processed food itself, the discipline of food science is a relatively recent phenomenon. The first department began in 1918 at the University of Massachusetts at Amherst with remarkably simple, practical food processing goals: to preserve fruit. Local farmers complained to the university's horticulture department that they were losing money on lower-quality fruits and other left-over product that couldn't be sold at markets. In response, the department head directed one of his botany professors, Walter Chenoweth, to set up a lab dedicated to the study of methods for fruit preservation. Chenoweth protested he knew nothing about the subject, but somehow this only made him seem more qualified for the job.

The practice of basic canning had been around a long time. Companies like Heinz, Campbell's, Borden, and Van Camp were successfully selling containers of jellies, horseradish, ketchup, condensed soups, and milk. Homemakers, too, used heated and airtight glass jars as a way to preserve nature's seasonal bounty. But the food shortages of World War I had sparked new interest in preservation, and there was still much to be learned about the chemistry of the process. Chenoweth spent the remaining years of his life acquiring this knowledge and sharing it with others. He hosted regular seminars for homemakers on how to apply scientific principles to their canning—how to avoid improper packing, prevent air pockets from forming, avoid sealing a jar too tightly, and apply just the right amount of heat.

Over the next four decades, several land-grant universities, which were set up by Congress with mandates to teach practical science, followed UMass into the food science future. Most of these programs were geared toward finding new ways to boost the consumption of crops grown within the state. Kansas State University focused on meat processing and wheat. The University of Wisconsin at Madison devoted itself to dairy. And the University of Illinois specialized in products from soybeans. Eventually these programs were expanded to include a broader range of studies, ultimately becoming more and more elaborate. By the fifties, it wasn't so much about food as "foodstuffs." One textbook in 1953 was entitled *Foodstuffs: Their Plasticity, Fluidity and Consistency.*

Loving Food

Notwithstanding the inverse relationship between food science and cooking, many of the Purdue students and former students I talked to said they were inspired to study food science explicitly because they like to prepare their own food. What about cooking school, I wondered? Nobody had seriously considered this, as most were driven by a greater affinity to science than art. Jenn Farrell, a senior from Indianapolis, described cooking as her "stress relief." Most evenings, when she returns to her apartment after working in the sensory lab, running the Food Science Club, and taking classes in ingredient technology and food analysis, she whips up meals from "whatever's lying around in the fridge." Chicken will go into the oven with a medley of vegetables and a sauce she's put together. Baking, she said, is her favorite activity. On weekends, she's likely to make banana bread from a coveted recipe handed down from her grandmother. "It's the best banana bread you've ever had," she crowed.

Farrell hadn't heard of food science until her high school organic chemistry teacher mentioned it to her during her junior year. The teacher suggested she might want to consider going to college to study something like chemical engineering, but that didn't interest her. When she mentioned her fascination with cooking and food, he gave her a newspaper article detailing the top five up-and-coming careers. Food science was one of them. The field immediately struck Farrell as an ideal way to leverage her passion for food into a career without having to deal with the mania and weird hours of restaurant life. "The next thing I knew, I was at Purdue taking Food Science 101," she said.

Farrell now works in Chicago for a company called Leahy IFP, a maker of canned fruit, boxed juices, pancake syrup, and margarita mixes. She got the job, a post in the research and development department, in the spring of her senior year and started working a few weeks after graduation.

Food science is booming. All of Purdue's 2012 graduates either were hired by the food industry or went on to further study. A poster in the hall of the food science building highlighted each of the companies that had offered full-time jobs or internships to students the year before, in 2011: Kellogg, Heinz, Cargill, Pepsi, Kraft, Morgan Foods, General Mills, Nestlé, Sara Lee, ConAgra, Maplehurst Bakeries, Sensient, Leprino Foods (cheese), Kerry Ingredients, McCain Foods, and AmeriQual (military meals). Suzanne Nielsen, the gracious, gray-haired, food science department head—a woman who could easily be mistaken for an English professor on her way over to teach a class on Chaucer—said that since the department was formed in 1982, they've seen 100 percent job placement for American students (those from other countries can have a harder time finding industry jobs, since employee work visas are a hassle companies would rather not deal with). This means that within six months of graduation, all those who wanted a job found one, even in years when the broader job market was in disarray.

It's no coincidence that the head of the department and the most recent president of the undergraduate Food Science Club are both female. The chemical and aeronautical engineering departments at Purdue and just about everywhere else remain boys' clubs, but food science has evolved to be predominately female. Nationally, women now account for 65 percent of students, though this was a long time in the making. Founded in 1939, IFT didn't have its first female president until 1997; fifty-seven men had previously occupied the job. The increasing numbers of women entering the field in the eighties caused then-president Theodore Labuza to worry that this influx of estrogen might "convert the industry into one which is female dominated, like nursing or teaching, and reduce the salary levels in the process." He wondered nervously, "Should more males be recruited into the field?"

International students, too, are overrepresented in food science, with roughly one fifth of those studying it coming from other countries, mostly China and India. Foreign food scientists are more likely to land jobs at universities and government agencies such as the Food and Drug Administration (FDA) and the U.S. Department of Agriculture (USDA). Others return to their home countries.

Purdue's heavily female, heavily international food science students don't just study how to make your hamburgers juicier and dips creamier. Food scientists do the important job of figuring out how to reduce harmful bacteria in food. They analyze the way food behaves in all kinds of situations and they measure all the stuff that's in it. They've looked at ways to increase vitamin D levels in mushrooms, at how microwave cooking affects the carotenoid content of chilaca chili peppers, and at what allergens are present in mangoes. Spend time with food scientists, and you'll learn all kinds of fascinating things about food: that you can make cranberry juice clear by changing its pH and that using fat-free dressing on a salad can prevent you from absorbing many of the vegetables' healthy (fat-soluble) phytochemicals.

At the same time, food science programs are charged with grooming students to engineer the next McDonald's snack wrap or design a new kind of prepackaged lunch. There's a mock cor-

porate boardroom at Purdue—to give students experience present-
ing to people seated around intimidatingly large, well-lacquered
tables. There's also a pilot plant where students can construct
foods using scaled-down versions of some of the same machines
that companies install in their factories. During an undergraduate food processing class I attended, the professor informed students they would be using the plant the following week to make
hot dogs. The announcement was greeted mostly with squeals of
delight, along with a few groans of disgust. "If you have a weak
stomach for meat and fat—because that's what hot dogs are—then
come talk to me," she advised.

"The industry was there at the table when we started," said
Philip Nelson, the department founder and building's namesake.
Nelson, who's in his seventies and was visiting campus from his
house on a lake in northern Michigan, recalled, "I hired a former executive vice president from Campbell's Soup as a mentor
to help me and to figure out how to work with food companies.
We established programs they said they needed, like carbohydrate
research." He called the food industry one of Purdue's "customers"
and said that serving its needs was on par with serving the needs
of students.

Many of the students I talked to acknowledged, with varying
degrees of candor, that it isn't always easy to endorse wholeheartedly an industry that churns out billions of dollars worth of Hot
Pockets, Doritos (some of which come from a plant twenty-five
miles from campus), Little Debbie snack cakes, sugary drinks,
and chicken wings that aren't wings. But like the food scientists
at IFT, they said that precooked and fast food is crucial in a hectic
world. By and large, they said they'd like to work toward replacing unhealthy ingredients such as sodium and unpronounceable
chemicals with healthier options like fiber and whole grains. Jenn

Farrell did an internship at a flavor company, where she helped soup and sauce manufacturers cut back sodium levels by using yeast extracts and hydrolyzed vegetable protein. Like Yan, a PhD student from Shanghai, told me she's been developing a new kind of starch that will digest more slowly, possibly serving as a replacement for white flour in bakery foods. "People are busy and there's a need for these products, but we can help people make better choices by providing healthier foods," she declared.

Creating healthier processed food is a noble goal. If future scientists like Yan and Farrell can engineer a way out—find salt, sugar, and fat replacements, use more nutritious ingredients—everybody wins. The processed food industry can grow its sales, while America shrinks its waistline and health-care expenditures. We can have our Little Debbies and eat them too.

But can we really?

It was technology, after all, and an unbridled belief in its utility for the food industry that helped get us into this mess in the first place. When applied to food, scientific innovation hasn't always been a good thing. And nobody knew that better than a Purdue chemistry professor who strolled the streets of West Lafayette some 130 years ago.

2

The Crusading Chemist

*When Americans think of consumer advocates, the names
Ralph Nader or Esther Peterson or Eliot Spitzer may jump to mind.
But Harvey W. Wiley, M.D., was the original.*

—The Food and Drug Administration

In the fall of 1902, just as the nation's first wave of urban migration tipped the population of our capital to 280,000, a dozen
men gathered in a basement along what is now Independence Avenue. They were dressed in dark suits and sat expectantly around
two dining tables covered in white linens. The meal they ate had
been prepared by trained chefs, served on elegant china—and laced
with poison. Everyone at the table was aware of this fact, but no
one knew that the specific poison was borax, a naturally occurring
mineral known to cause skin reactions, respiratory irritation, and
various forms of gastrointestinal distress.

The meal passed uneventfully; no one keeled over. The twelve men showed up the next morning for another borax-laced dining experience. They did this every day for nine months, getting three free meals a day at the Bureau of Chemistry's basement kitchen. When they weren't eating, the men, all of whom had volunteered for this ordeal, submitted to meticulous recordings of their weight, temperature, and pulse rate, as well as regular collection of their urine and feces. One participant commented that the hardest part of the experience was not the consumption of hazardous food but figuring out how to get all of his bowel movements turned in for analysis.

This unusual human-guinea-pig experiment—the likes of which today's more morally and legally constrained scientists could only dream of—was aimed at testing seven new and worrisome food additives: salicylic acid, sulfuric acid, copper sulfate, potassium nitrate, sodium benzoate, and formaldehyde, along with borax. It was the brainchild of a chemist who would go on to become the founder of modern food regulation. Harvey Wiley, a tall man with a sweeping forehead, large bulbous nose, and a head that cocked forward, had come from Purdue University in 1883—long before it had something called a food science department—to work for the Agriculture Department during a profoundly transitional period in U.S. history. After growing their own food for more than a century, Americans had begun to rely on companies, often in faraway cities, to produce their food. With no federal laws governing this incipient modern production, all sorts of shady things were taking place. A deeply principled man, Wiley was the first to want to do something about it.

Although nobody died during eight different rounds of what came to be known as the Poison Squad, many volunteers—who were never in short supply, thanks to the considerable charisma of the man they took to calling "Old Borax"—complained of nausea, vomiting, stomachaches, headaches, and the inability to perform work of any kind. When someone's symptoms became severe, Wiley would put them on clean food for several weeks before they returned to the study. Through meticulously detailed reports, he argued that each of the tested substances was not fit for repeated human consumption, even in small doses. But none of the volunteers—who were young and healthy and endured their poison assault for no more than a year—suffered lasting harm. One of the last surviving members of the Poison Squad, William Robinson of Falls Church, Virginia, lived to the advanced age of ninety-four before passing away in 1979.

A Pure Food Cause

The man who would become the nation's first nutrition champion and food industry critic was born in a log farmhouse in 1844 in the tiny town of Kent, Indiana, one of seven children raised by Scottish-Irish immigrants. The family lived on a farm and the land provided them with wool, wheat, corn, and fresh game. Although it was their livelihood, Wiley's father, who fancied himself an intellectual, never really took to farming. He taught himself Greek and, at a time when very few women studied past the sixth grade, Mr. Wiley encouraged his daughters to attend college for some span of time. In a state that had banned slavery some forty-five years before the Emancipation Proclamation, the family embraced abolitionism. Every Friday night Harvey and his siblings gathered around a fire to listen to their dad read the *National Era,* an abolitionist magazine that serialized *Uncle Tom's Cabin* and published popular authors like Nathaniel Hawthorne. In the 1850s, the Wiley farm on the wooded Indiana hills became the southernmost station of the Underground Railroad, a post where escaped slaves were ferried from the nearby Ohio River to a station eight miles north.

Young Harvey Wiley dreamed of becoming a doctor, a flight from farm life his father wholeheartedly supported. At eighteen, he enrolled at Indiana's first private university, Hanover College. Afterwards he served for a brief time as a Union soldier in the Civil War and then went on to study at Harvard. After his schooling, he accepted a series of lucrative teaching positions, including the one in the chemistry department at Purdue. He figured that after a spell he would leave teaching to become a doctor. But one day, while visiting a lab in Berlin, Germany, he looked through a new device that would change the course of his life. Called a polariscope, it was a cylinder with crystal prisms at each end. It could measure for the first time the individual constituents of certain foods. Wiley was fascinated.

Back in Indiana, he persuaded the university to buy one. He started analyzing food samples, and the results were stunning. He looked first at a variety of syrups—honey, molasses, "pure Vermont maple syrup," and sugar derived from sorghum. The polariscope revealed that nearly all of it contained a good deal of cheap glucose derived from corn. He then turned his attention to bottles of "imported Italian olive oil." They proved to be nothing more than Alabama cottonseed oil. Strawberry and raspberry jams were concocted from rotting apples and pulp, along with glucose and chemical flavorings derived from coal. Tomatoes had been "preserved" with salicylic acid. Bread was bulked up with sawdust. The more he looked, the worse the situation seemed and the more concerned he became. As the sort of man who saw right and wrong in sharp contrast, Wiley regarded these vast misrepresentations as morally reprehensible. As he saw it, customers should be able to know what they were buying. This sentiment would become one of the central principles of his life's work.

With no legal restrictions on ingredients or labeling disclosure in place, American homemakers had no way of knowing what they were truly serving their families and whether it was harmful. What was needed, Wiley concluded, were national rules governing what could and couldn't go into our increasingly industrialized and centralized food supply—laws that would establish the importance of "pure food," as he called it.

A half-century earlier, nearly every American lived on a farm, either producing everything they ate or trading with neighbors. By 1900, roughly half of the American population had left their agrarian outposts in favor of city living. To feed these urban masses, food had to be preserved and production had to be centralized. Kansas City and Buffalo joined Minneapolis as hubs for the milling industry, and much of the meat supply came out of Chicago. Canning operations started popping up all along the East Coast. While not all these manufacturers—at this point still small, modest enterprises—were unscrupulous, those who used chemicals and other new technologies as a way of cutting costs often gained a competitive edge over those making the real thing, forcing the honest sellers either out of business or into a reluctant embrace of cheaper production methods.

Wiley's report to the Indiana Board of Health on the sugar and syrup adulterations he had found garnered attention in agriculture circles, and he was invited to a convention of sorghum (a grass cultivated for sugar) growers in St. Louis. There he met and impressed George Loring, the head of the newly formed federal Agriculture Department. Several months later, Loring offered him the position of the department's chief chemist, and Wiley took it, derailing whatever remaining plans he still had of practicing medicine. Ultimately, he felt he could make greater strides in safeguarding the health of his fellow Americans by being in Washington than he could as a doctor in Indiana. And nothing was more crucial to this goal than the passage of federal pure food legislation. Never did he imagine, though, that it would be nearly a quarter of a century after his arrival in Washington before such a law was finally passed.

Not that there weren't many attempts. In 1889, Senator Algernon Paddock of Nebraska, working with Wiley, introduced the first pure food and drug bill in Congress. It narrowly passed in the Senate, only to be defeated in the House, a humbling pattern that would be replicated more than half a dozen times. Sometimes the bill won passage in the House, only to die in the Senate. Although Wiley had a few key allies in Congress, much of the food industry was adamantly opposed to the notion of government interference. So, too, were the liquor industry and drug makers, which the bill was also trying to regulate. "Poor mothers doped their babies into insensibility at night with soothing syrups containing opium or morphine," Wiley wrote in his autobiography. In 1902, an official from a large food distribution company told Congress that Wiley's bill would single-handedly ruin all sectors of the food industry. "Make us leave preservatives and coloring matters out of our food, and [make us] call our products by their right name and you will bankrupt every food industry in the country," he warned.

Wiley found himself the subject of much criticism—"a veritable fusillade of poisoned arrows," he would later say. "There was hardly a week that some interested organization or mercenary interest did not demand my removal from public service." He also claimed that detectives had been hired to root around in his personal life. Not that there was much to report. Thoroughly immersed in his work, Wiley never managed to date much or to marry, though he once proposed to a considerably younger woman, who turned him down.

Instead of growing disheartened by the attacks, Wiley hardened his resolve. He acquired additional supporters in Congress and engaged leaders of women's groups, such as Alice Lakey of the General Federation of Women's Clubs. In 1901, Theodore Roosevelt ascended to the presidency under the tragic circumstances of the assassination of William McKinley. Though he grew to dislike Wiley personally—both men were equally obstinate and opinionated—Roosevelt had enormous respect for his chief chemist's abilities and integrity. And the twenty-sixth president understood the necessity of having the federal government responsible for ensuring a safe and healthy food supply. In 1898, while Roosevelt was leading a regimen of cavalry in Cuba, hundreds of American troops sent over to the island were sickened after eating spoiled canned meat shipped from the United States.

Wiley knew that legislative approval hinged on broad public support, and his Poison Squads proved invaluable on this front. They were intended as sober scientific analysis, but journalists found them fascinating fodder for splashy stories. After Wiley discovered that writers had been interviewing his chefs through the basement window, he decided to welcome them in to show them the venom he was serving up—and to let them know that most Americans ate some of the same concoctions in their food every day, though in smaller concentrations. By 1903, just about everyone knew what Wiley was up to. A minstrel show that year featured a musical number called the "Song of the Poison Squad."

It wasn't until publication of Upton Sinclair's *The Jungle*, however, in early 1906 that Wiley's crusade hit a tipping point. The wildly popular and influential novel detailed the deplorable conditions and inhumane, corrupt working environment at the nation's meat stockyards. Based on Sinclair's seven-week stint at meatpacking plants in Chicago, it included accounts of spoiled meat being doctored in secret and workers falling into rendering tanks and being ground, along with animal parts, into "Durham's Pure Leaf Lard." Such details were reminiscent of Wiley's descriptions of aging, rotting meat treated with chemical preservatives and sold as fresh and of eggs that were deodorized with formaldehyde and sold for cake making. "I aimed at the public's heart," Sinclair pronounced, "and by accident I hit it in the stomach." Public outcry was swift, and by June of 1906, President Theodore Roosevelt had signed the nation's first laws governing food—the Federal Meat Inspection Act and the Pure Food and Drug Act.

Food Fights

As is so often the case with pioneering legislation, the laws represented a deeply compromised version of what their most tireless advocate had hoped for. Wiley had wanted certain substances—the ones he tested on the Poison Squads, for starters—to be outlawed. He wanted other additives and nontraditional ingredients listed clearly on food packages. But he didn't get any of that. The final version of the bill gave the government the power to forbid the manufacture of any food that was adulterated, misbranded, or deceptive. But its language was vague about what that meant, as were rules about which chemicals could and couldn't be added to food.

It was perhaps inevitable that the purity of Wiley's vision would be watered down by the political process and lobbying efforts. The legislation in the summer of 1906 didn't mark the end of Wiley's crusade for better food—it was the beginning. He would spend the remaining years of his government career waging heated battles against ingredients and products he considered to be misleading or harmful. More often than not, these were battles he lost.

One of his first actions concerned corn sweeteners, substances still causing controversy more than a century later. In 1902, the Corn Products Refining Company (now Ingredion) started selling its glucose syrup directly to consumers, renaming it "corn syrup" because there was a widespread misperception that glucose was somehow derived from glue. Wiley thought the term "corn syrup" should be reserved for juice that came directly from corn stalks—something that had been made by natives in North America for at least four hundred years. He argued for changing the name back to "glucose." His boss, Secretary of Agriculture James Wilson, however, disagreed and gave Corn Products sanction to sell its glucose as Karo Corn Syrup, a name it's still known by today.

Wiley also lost on saccharin, the world's first artificial sweetener and the inaugural product for the Monsanto Corporation. Canned-food manufacturers were using it as a cheap alternative to sugar, something Wiley considered to be both deceitful—lessening a product's quality—and potentially unsafe, with the potential to cause harm to the kidneys. But President Roosevelt, whose doctor had been prescribing the sweetener to him for weight control for several years, personally intervened. "Anybody who says saccharin is injurious to health is an idiot," he bellowed during a meeting, fists clenched and face "purple with anger," as Wiley recalled it.

As time went on, the portly, bespectacled president lost confidence in Wiley's ability to back up his sometimes grandiose claims. Wiley continued to lock horns with food manufacturers, who demanded to know why exactly they couldn't keep using certain additives. To lend more credibility to government decisions on food additives, Roosevelt appointed a five-man board to review Wiley's Poison Squad data. The board eventually agreed with Wiley on five of the seven substances, banning borax, salicylic acid, sulfuric acid, copper sulfate, and formaldehyde from American food. Potassium nitrate was declared legal. The fate of the seventh one, sodium benzoate—one of those white powders in Subway's teriyaki glaze and fat-free sweet onion sauce—remained contentious because ketchup manufacturers argued it was essential for preventing spoilage. Wiley had once believed that small amounts of this preservative were acceptable in food, provided it was declared on the label, but when his Poison Squad experiments showed otherwise, he reversed course. Sodium benzoate was, he wrote, "highly objectionable and produces a very serious disturbance of the metabolic functions, attended with injury to digestion and health."

Wiley found a rare food industry friend in Henry John Heinz, who had figured out a way to stop using sodium benzoate for ketchup, starting with higher-quality, fully ripe tomatoes and sterilizing the bottles at the factory. Breaking ranks with other manufacturers, he highlighted the new method in the company's marketing. In 1909, a Heinz advertising campaign characterized sodium benzoate as a chemical for lesser ketchups, something that masked unsanitary handling, lazy manufacturing, and inferior raw materials. Ketchup with this preservative, ads stated, was "the kind of food you would not care to eat if you could see it made and what it is made of." Another read: "Good ketchup needs no drugs." Despite Heinz's enthusiastic support, Wiley's efforts to ban sodium benzoate were fruitless. In 1911, a conference of state food commissioners, Department of Agriculture officials, physicians, and scientists narrowly voted to allow the continued use of sodium benzoate in food, provided the label indicated its presence.

Wiley's final attempt to regulate potentially harmful products was a bold legal undertaking against the increasingly popular drink Coca-Cola. Ever since Coke's introduction in 1886, Wiley had harbored misgivings about the beverage, which contained caffeine and was heavily consumed by children. "Habit-forming and nerve-racking," he called it. At Wiley's behest, government agents set up a stakeout on the Tennessee-Georgia state line early one October evening in 1909 to wait for a truck traveling from Coca-Cola's plant in Atlanta to a bottling factory in Chattanooga. When the truck arrived, agents seized forty barrels and twenty kegs of syrup. The government used what they found as grounds for a lawsuit accusing Coca-Cola of misbranding, since the drink contained an adulterant (caffeine), no coca, and little if any trace of nuts from the kola tree. The drink had once contained a cocaine-type substance from the coca plant, but by 1902 most of that had been removed amid growing concerns about the drug's addictive nature.

When the case went to trial in 1911, a federal court in Tennessee ruled that the government had failed to show evidence that anyone had been harmed by consuming caffeine or that there was any law against selling it. This decision was later overturned by the Supreme Court and sent back to the original judge in Tennessee. In the end, Coke negotiated an out-of-court settlement with the government, agreeing to reduce the amount of caffeine by half and add flavorings from the coca leaf and kola nuts. Coca-Cola, of course, went on to become the world's largest soda manufacturer.

Wiley fared less well. Attorney General George Wickersham and one of his staff members had come to regard Wiley as an industry-hostile zealot, and they used the Coca-Cola affair as an opportunity to end his twenty-nine-year career in government. They accused the chemist of paying an expert witness in the Coke trial excessive amounts of money, constituting bribery. Though an investigation in the House of Representatives cleared Wiley of the charges, the public spat, coupled with an Agriculture Secretary who'd grown "alertly antagonistic" to Pure Food and Drug Law enforcement, convinced Wiley that he could no longer be effective in his position. In 1912, he resigned from the Bureau of Chemistry. One newspaper headline declared: "Women Weep as Watchdog of the Kitchen Quits after 29 Years."

An Unsuited Hero

In the years following Wiley's departure from government, Americans were treated to a variety of new packaged, ready-to-eat foods. Oreos debuted in 1912. A year later, Mallomars arrived, and in 1915, Kraft processed cheese. Four years after that, Americans had Hostess Cup Cakes, followed by Good Humor ice cream, and then, in 1921, that quintessentially American food, Wonder bread. Now on the outside looking in, Wiley believed more than ever that American eaters needed an outspoken advocate for their nutritional health, and he took a job with *Good Housekeeping* magazine as its resident health and consumer products expert. For Wiley, it was a powerful platform. He had veto power over the magazine's food and drug ads and he estimated that he rejected more than a million dollars' worth of advertising revenue over the course of seventeen years, which amounted to a fortune in those days. He also helped implement the magazine's famed seal of approval, sending food and other items to labs for analysis and awarding ratings of "approved," "noncommittal," or "disapproved."

Wiley's writing appeared regularly in the magazine. In one piece, he condemned new forms of ice cream appearing on the market, products he didn't consider to be ice cream at all since some of them contained no actual cream. Instead they harbored lots of skim milk and thickeners such as gelatin. "Gelatin is often made from materials that are themselves inedible," he wrote. "Sometimes it is made in glue factories, and no one can tell just where the glue ends and the gelatin begins." In the same column, Wiley weighed in on the use of food dyes that were permitted to contain small amounts of arsenic. He wrote:

> One of the unfortunate tastes which has been cultivated in this country is a desire to eat highly colored ice cream. The dangers from these colors in ice cream is not that of serious illness or death, but of a continued attack upon the vital centers [a term Wiley used to mean organs]. . . . My advice is to refuse to eat any foods colored with aniline dyes. Red colors can easily be secured from small berries and other fruits and these should be the only ones tolerated.

Divisive at the time, many of Wiley's rants were prescient. The advice, for instance, that if people were going to drink sodas at all they should do so slowly, allowing time for the satisfaction of thirst to register so that too much of the highly sweetened liquids wouldn't be consumed, presaged our current understanding of satiety signal mechanisms. In 1927, long before America's health establishment rejected cigarettes, Wiley expressed suspicion that the use of tobacco might be harmful and could promote cancer. And his estimation of white flour, with all its nutrient-rich bran and germ removed, as "the base of nearly all the bad nutrition in the United States" jibes with what we know to be true today.

At the time, though, there were medical experts who regarded whole-wheat bread as difficult to digest and therefore less nutritionally desirable than white varieties. Even as late as 1930, the American Medical Association (AMA) held white bread in at least equal regard to whole wheat, calling it a "wholesome, nutritious food with a rightful place in the normal diet of the normal individual." It took another decade for the AMA and other experts to get their facts straight. By 1941, both the doctors' group and the government were urging the milling and baking industries to enrich their refined grains with the important vitamins that had been lost in processing.

Wiley's view of health and nutrition wasn't the only way in which he evinced preternaturally contemporary instincts. The *Good Housekeeping* gig afforded him considerable freedom to go on speaking tours and "spend time with his family." Shortly before leaving the Bureau of Chemistry, he got married for the first time at the age of sixty-seven to the younger woman he had first proposed to over a decade earlier. They raised two sons and enjoyed nearly twenty years of marriage, spending a good deal of it on their farm in Virginia. After a busy, productive career in Washington, DC, he acknowledged, as increasing numbers have today, a strong desire to explore local food production. He raised dairy cows on his farm and in 1915 wrote a book called *Lure of the Land: Farming after Fifty*. When he died on June 30, 1930, Wiley was eighty-six, some twenty-eight years past the average male life expectancy at the time. His death came, poetically, on the same day and month when President Roosevelt had signed the Pure Food and Drug Act into law twenty-four years earlier.

Today, the FDA's Harvey W. Wiley Federal Building in College Park, Maryland is a long, flat building with a checkerboard of windows. It has a sharply angled roof that brings to mind a paper airplane. Inside the lobby, along a corridor leading toward an auditorium, an exhibit describes the life and work of the man for whom the building is named. Black-and-white photos of Wiley line the wall, with captions explaining how this one man, more than all others, paved the way for all food regulation that was to come.

Were he alive today, Wiley would certainly appreciate that the government now has an agency in charge of maintaining the "safety, healthfulness and honest labeling" of our food supply; the FDA wasn't actually formed until the year he died. On the other hand, he'd likely find fault with much of what's taken place under the agency's watch. Profiles of Wiley, like the one spanning the lobby wall and those found on the FDA's Web site, tend to highlight his efforts to rid the food supply of rotting food and bad actors while omitting his broad-based criticisms of highly processed, additive-laden food. In one of his last speeches as chief chemist, Wiley summed up his beliefs with biblical overtones: "Give him that buys that which he purchases, and add not thereto anything which conceals damage or inferiority. Mix no drugs with food. Render unto the green grocer's and unto the pharmacist that which is his own." And then there was this simpler declaration in a 1914 *Good Housekeeping* publication: "I have always stood for food that is food."

I asked Suzanne Junod, an FDA historian who's written about Wiley, what the agency's spiritual progenitor would likely think about the current food landscape. She knew right away what I was talking about. "I think that going into a supermarket today would throw him for a complete loop," she said. "I think he'd be astounded at Cheetos and definitely be surprised at how much soft drinks, which he used to call 'medicated soft drinks,' have withstood the test of time." It's not hard to imagine that Wiley would also be stunned by both the sheer number of food additives in use today and the fact that some of them have simply been declared safe by the companies making them without so much as a glance from the FDA. As someone who believed in honest representation above all else, he would probably be horrified by the decision two decades ago by FDA officials, some of them working in the Harvey Wiley building, to fling open the supermarket doors to genetically modified (GM) foods without labeling regulations that would enable us to identify them.

Wiley's most crushing disappointment might be the fact that a hundred years later, we're still consuming his old bêtes noires—sodium benzoate and saccharin. Banned for use in food in 1911 and then un-banned during World War I, saccharin has had a roller-coaster record of safety. Studies done in the early seventies linked the chemical sweetener to bladder cancer in rats, prompting the FDA to attempt to ban it again. But outcry from the beverage industry and millions of diet soda drinkers forced the agency to shelve this plan. Instead they required all products containing saccharin to carry warning labels stating that the substance is a proven animal carcinogen. Three decades later, when scientists discovered that the urine of rats is different from that of humans in ways that make their bladders more susceptible to cancer when saccharin is consumed, the FDA repealed the warning label requirement. In 2000, they declared the 130-year-old chemical safe.

Sodium benzoate, meanwhile, a preservative made from a petrochemical found in paint thinner, continues to be used in a variety of foods—condiments, salad dressings, sauces, frozen foods, fast-food meals, and soda, including Mountain Dew, Sprite, Dr Pepper, and Mug Root Beer. Heinz, once Wiley's loyal ally, uses it to ward off bacteria in the company's Heinz 57 Sauce, and the buttermilk ranch dressing it makes for KFC, though the company never did add it back into ketchup. Wiley's assertions about sodium benzoate—drawn from his Poison Squad trials—that the chemical causes a "disturbance of the metabolic functions" and "injury to digestion" have never been scientifically proven. Yet it remains a controversial ingredient, implicated in childhood hyperactivity when used with artificial food colorings.

For food manufacturers wishing to steer clear of contentious additives, there are a dozen other preservatives available to help keep food decay-free for long chunks of time—potentially for the next hundred years.

3

The Quest for Eternal Cheese

Age is something that doesn't matter, unless you are a cheese.

—Billie Burke, American actress

As Wiley's career in government was skidding toward its final days, the ambitious son of a Canadian dairy farmer began work on what would become one of America's most enduring modern foods. James Lewis Kraft's life is a classic tale of twentieth-century determination and entrepreneurship, rooted in humble small-town beginnings and culminating in the creation of a fifty-four-billion-dollar food company. His innovation tapped into a changing food ethos among American housewives and served as a model for processed food to come—products that were attractively packaged, nationally advertised, longer lasting, more convenient, and of inferior nutritional value.

James Kraft grew up one of eleven children in Stevensville, Ontario, a farming community just south of Niagara Falls. Much of his free time was spent working on his family's dairy farm, but like Wiley, he sensed that his fate lay elsewhere. After working for several years as a clerk in a general store in neighboring Fort Erie, he got a job across the border in Buffalo for one of the cheese companies that had supplied his store. His real ambition, however, was to start his own business, and in 1903, at the age of twenty-nine, he moved to Chicago with $65 in his pocket. He used his money to rent a wagon and a horse named Paddy, and before long, Kraft and Paddy could be seen arriving every morning at Chicago's biggest cheese market, on South Water Street.

A small man with searching brown eyes and a crooked, wry smile that people often found endearing, Kraft always arrived early to snatch up the best cheese at the market. He spent the rest of the day clip-clopping around to the city's grocers. Many of them had been going to the trouble of fetching their own cheese and were happy to pay a little extra for someone to both select and transport it. Smart and driven, Kraft had a knack for understanding what his customers wanted, sometimes before they knew it themselves.

He knew with certainty that customers would want a cheese that didn't go bad. Refrigeration technology was still in its infancy, and few food stores had anything more than a few blocks of ice to keep things cool. Grocers took losses every time cheese went moldy or started to separate, which could occur after just a few weeks, especially in the summer. Cheese selling was inefficient. In the days before the wonders of Saran Wrap and plastic vacuum packaging, stores received their cheddar and Swiss in wheels as large as car tires. Each morning grocers had to slice off the dried-out surface of the wheel that had hardened overnight. Kraft figured if he could find a way for cheese to be delivered in smaller, long-lasting packages, grocers could avoid this waste. If he could do this, he might have the makings of a blockbuster product on his hands.

One evening, on his way home from a day of peddling cheese, he bought a copper kettle. In his apartment, he filled it up with grated cheddar and heated it, attempting to kill off the bacteria responsible for both aging cheese and turning it putrid. But all he got was a sticky, oily mess. Unfazed, he kept at it, eventually setting up a makeshift laboratory. In 1914, World War I broke out. The Armed Forces in Europe needed food that would keep under any condition, and Kraft intensified his efforts. After many attempts, he lit upon a breakthrough one afternoon after absentmindedly stirring the heated cheese longer than usual, about fifteen minutes. The extra mixing homogenized the cheese solids with the fat, so it didn't separate. He'd created a smooth, creamy glop of cheddar thoroughly free from troublesome bacteria.

The whole thing was a relatively simple process, yet one that would transform the entire concept of cheese. As Kraft wrote in the patent he received in June 1916, this new product could be "kept indefinitely without spoiling." Kraft began selling his eternal cheese in 3-ounce and 7-ounce aluminum tin cans, ultimately producing six million pounds of it for the U.S. military. And so in France, the nation with the world's proudest cheese heritage, many Americans got their first taste of processed cheese.

Back in Chicago, the new cheese became an immediate hit. Although the product was more expensive, people loved it because it was dependable. It had the same soft, velvety texture and mild taste every time. In contrast, the production of natural cheese was wildly inconsistent due to variances in live cultures and the sheer number of producers making it. Sometimes housewives got cheddar that was creamy; other times it was crumbly and granular. To families without modern refrigeration, processed cheese was a marvel. You could stock up the pantry with Kraft's glossy tins and keep them for weeks, months, or even years.

As the J. L. Kraft & Bros. Company (J. L. was joined by four of his ten siblings—Norman, John, Fred, and Charles) expanded into new markets, it launched the first national advertising campaign for cheese, turning a nameless commodity into a recognizable and trusted brand. Ads stressed the superiority of Kraft's product, which was then called "pasteurized loaf cheese," and underscored its link to technological modernity. One ad in 1924 boasted, "Kraft cheese is not 'just cheese' in a newfangled package. It is modern ideas and modern methods applied to cheese making." While today many eaters champion local products and local ingredients, Kraft actually trumpeted the fact that one of his new cheese factories drew milk from as far as thirty-five miles away. In an era when most cheese makers purchased their milk from farmers living within two miles, Kraft's huge footprint was a novelty and a selling point.

"Cheese in tins is the new, safe and clean way to buy cheese," another ad stated. "Old unsanitary methods of marketing this sensitive food are fast going the way of the oatmeal bin and cracker barrel." Ironically, when Kraft Foods, as it was renamed, began packaging a new brand of natural cheese in 1954, a year after J. L. died, it decided to invoke those very same old-school technologies, choosing the name Cracker Barrel.

The ads helped propel the J. L. Kraft & Bros. Company from $500,000 in sales in 1917 to $36 million in 1926. Although many housewives during this era were still wary about feeding their families from cans, they were intrigued by scientific advances taking place in food production and the reliability and convenience they promised. The broad and relatively swift acceptance of processed cheese signaled a shift in consumer sentiment. By 1930, processed cheese represented more than 40 percent of all cheese consumed in the United States, though by this point Kraft had competitors.

Through its marketing, the food industry would continue to normalize food's newfound alliance with technology.

Not surprisingly, Kraft's sweeping success was deeply troubling to makers of natural cheese, who were often prevented from entering the processed-cheese market because of patents. They regarded Kraft's product as inferior in taste and quality and unworthy of being called cheese. A Chicago dealer who spoke at a 1922 convention called it a "dead mass." Officials in Wisconsin, the American epicenter of all things cheese, were particularly anxious, and in 1925 the state legislature called for an investigation into the processed-cheese industry. What the Wisconsin Dairy and Food Commissioner found was that Kraft and other manufacturers often used low grades of cheese that weren't properly aged. The Roquefort, Swiss, and Camembert versions of processed cheese all contained large amounts of the same thing—mild cheddar. Most processed cheeses, the commissioner said, also contained more water and less butterfat than required by state guidelines.

Some Wisconsin cheese makers went so far as to insist that these packages be labeled "embalmed cheese" or "renovated cheese." Fortunately for Kraft, that never happened. When the FDA established guidelines several years later for the production of these new cheeses, it chose the less horrific moniker "process cheese."

Sellers of traditional cheese held fast to their beliefs that natural, aged, nonsterilized varieties offered a superior taste. But in 1932, University of Wisconsin researchers published a study revealing that most Americans didn't necessarily agree. In blind taste tests, two thirds of people preferred the flavor of the processed variety.

Today, American cheese, as we now call it, continues to be immensely popular. It persists as the standard topping for fast-food burgers and sandwiches, a result of its low cost and refusal to become oily when melted. Even supposedly upscale burger chains like Five Guys use processed cheese exclusively on their cheeseburgers. As a consequence, we're scarfing it down in greater quantities than J. L. Kraft's day—now 7 pounds per person per year on average, though this includes Cheez Whiz (introduced in 1953), cheese spread, and other processed concoctions.

Over the years, the making of processed cheese changed remarkably little. Sodium phosphate, which a German company had just started manufacturing, was added in the twenties to help hold the mixture together. In the forties, Norman Kraft developed the roller technology for making slices, giving the company a huge boost in sales. The company has since refined J. L.'s original formula in minor ways, experimenting with different emulsifying salts, preservatives, and acidity regulators. It wasn't until the nineties, however, that Kraft scientists figured out a way to replace a portion of the original, core ingredient—the natural cheese—with a cheaper, more convenient milk-like substance.

Milk Cracking

Back when J. L. Kraft dreamed up a sterilized cheese, milk was just milk. For quite a long time, pretty much all you could do with it was make cheese, butter, cream, or yogurt. In the fifties, spray drying of milk became widespread. Then in the eighties, dairy got even more interesting. Companies such as Koch Membrane Systems, owned by the politically active Koch brothers, developed highly sophisticated membranes made from a type of plastic called polyethersulfone that are able to withstand heat and bleach-cleaning treatments. These membranes, known as ultrafiltration and microfiltration technology, could effectively take milk apart, separating it by molecular size into fractions. Much as crude oil is cracked into its various hydrocarbon molecules, milk could now be processed into a whole new category of processed food ingredients. Milk protein concentrate would help thicken yogurt and cut down on the amounts of cheese needed in boxed macaroni and cheese; whey protein concentrate allows for low-fat ice cream and half-and-half; casein and caseinate help create imitation cheese for low-priced frozen pizzas and give fat-free lusciousness to coffee creamers and whipped toppings.

Sometime in the mid-nineties, Kraft noticed that milk protein concentrate, then quite inexpensive, could be used to help trim costs. If the company replaced just a quarter of the cheese in its Kraft Singles with this ingredient, it could both cut costs and yield a more consistent product. Protein levels in cheese can vary considerably and the ability to add in varying amounts of milk protein concentrate as needed ensured precisely the same finished product every time. Furthermore, milk protein concentrate was easier to transport than cheese and could be stored for at least two years. Kraft's scientists retooled their formula to incorporate this durable milk-esque ingredient, effectively making their cheese even more processed than it already was. Profits rose, and the switch was an unqualified success.

Then the FDA got wind of it. During routine inspections of three Kraft facilities, the agency noticed that Kraft was now using milk protein concentrate. In late 2002, the agency sent a letter to Betsy Holden, Kraft's CEO at the time, stating that rules for processed cheese (one of those two hundred eighty foods for which the agency has standards) don't allow for the use of this ingredient. The company's cheddar and Swiss slices were in violation of the law.

This left Kraft with several options. It could revert back to its earlier recipe. It could lobby for an amendment of the processed-cheese standard. Or it could pursue a third, infinitely simpler solution—change the description of the product, which appears on the front of Kraft Singles packages in small letters. Doing this would take Singles conveniently out of the purview of FDA rules. So in 2003, America's favorite cheese slices were changed to read "Pasteurized Prepared Cheese *Product*" instead of "Pasteurized Process Cheese *Food*." The lettering was minute and the taste difference imperceptible. Most customers never noticed that their cheese had been downgraded from a "food" to a "product."

The various companies that make cheese slices for chains like McDonald's and Burger King were tempted to amend their own processed-cheese recipes. But because most fast-food menus appear in large print, they never actually did. At McDonald's, it might have meant that double cheeseburgers, one of the chain's most successful and popular items, would have to be recast as *Double Cheese-Product Burgers*.

Kraft continues to use milk protein concentrate in its Singles even though much of the cost advantage has been eroded due to higher prices for this ingredient. It still helps ensure consistency.

Is It Cheese?

Even when a fractured ingredient like milk protein concentrate is used, processed cheese is still largely a dairy product, with just 3 percent of its weight coming from additives like emulsifiers and preservatives. Notwithstanding its rubbery texture and Day-Glo sheen, many scientists who work in the dairy industry consider processed cheese to be cheese, plain and simple. When you go into Applebee's and order a hamburger, nobody asks you if you want a slice of *pasteurized cheese food* on it. It's just called cheese.

So is J. L. Kraft's innovation really all that different from the countless varieties of cheese that have been around since at least the Roman Empire? The answer appears to be yes, on a number of counts. For one thing, its fat content is often lower. The regular version of Kraft Singles has 32 percent less fat than Cabot Vermont Cheddar, for instance. This was once considered an unqualified benefit when nutritional guidelines demonizing all fats were in vogue. The current view on fats, however, is more nuanced, as some types of fat, perhaps even some saturated ones, are acknowledged to be beneficial.

Processed cheese also has more sodium, which Americans are vastly overconsuming. Excess sodium intake can lead to high blood pressure, one of those chronic conditions so many of us suffer from. A slice of Kraft Singles contains twice the sodium of a similar serving of Cabot Cheddar, due in part to additives like sodium citrate (an acidity regulator) and sodium phosphate (an emulsifier). Neither adds flavor or nutrition.

The use of sodium-based ingredients—there are dozens in our food supply—is part of the reason the lion's share of sodium we ingest comes from processed foods, not the saltshaker. Besides sodium citrate and sodium phosphate, there's sodium acetate, sodium acid pyrophosphate, sodium alginate, sodium caseinate, sodium hexametaphosphate, sodium nitrate, sodium stearoyl lactylate, and Harvey Wiley's favorite, sodium benzoate, to name a few. The presence of these substances in foods like soups, breads, salad dressings, dips, and fast-food sandwiches can go largely undetected, delivering unhealthy heaps of sodium without making the foods taste all that salty. Processed cheese, too, can be similarly deceptive, sometimes tasting less salty than natural varieties with lower sodium.

When consumed in excess, sodium-based additives are troubling because they bypass the body's ability to detect and regulate sodium intake. Oversalt your food at home, and it's unlikely you'll be able to remain unaware. Probably you'll end up finding something else to eat.

The other difference between processed and natural cheese is a distinction much less obvious, entailing something nutrition researchers and scientists are only beginning to understand, namely beneficial bacteria. When cheese ages over a period of months, its original starter cultures die off and are replaced by other types of bacteria. The longer a cheese ages, the greater the number of living organisms taking up residence within it. These bacteria are the reason people who are lactose intolerant can sometimes eat cheese incident-free. The bacteria feed on lactose, breaking it down so we don't have to. Processed cheese, in contrast, doesn't have any bacteria since killing them off—and thus extending the product's shelf life—is the whole idea. But this eradication may not, in the long run, be nutritionally advantageous.

We tend to hear the word *bacteria* and think of illness, of something nefarious that needs to be drowned in a generous pump of antibacterial soap. But the microbes found in aged cheese are not only totally harmless, they might also be good for us, contributing to the vast ecosystem of microorganisms in our guts and promoting healthy digestion much in the way the bacteria in yogurt and other fermented foods do. Mind-boggling and little understood, the gut microbiota, as it's called, is considered by many researchers to be an essential organ on par with the brain, affecting not only digestion but energy metabolism, weight gain, and the ability to ward off infections. There are so many tiny lives inhabiting our gastrointestinal tracts that they outnumber the cells in our entire body and contain a hundred times more genes than the human genome.

Efforts to document the benefits of fermented foods like cheese are in the early stages. Dean Sommer, a food technologist at the Center for Dairy Research in Madison, Wisconsin, has shown that the levels of bacteria in a gram of cheese range from 1 million to 100 million. (Comparatively, a gram of yogurt contains at least 100 million organisms.) Some of these cheese bacteria are certain to be quite beneficial, Sommer says; we just don't know yet which ones or how they function.

There's also the distinct possibility that a particular class of bacteria commonly present in cheese serves to break down some of cheese's casein protein into what are known as bioactive peptides, substances thought to ward off infection and stabilize blood pressure.

Aside from cheese being stripped of potentially healthy bacteria, there is the issue of American cheese's eerily long life span. American cheese doesn't quite last forever, as J. L. Kraft promised— its official shelf life, as marked on packages, is six months—but it can persist seemingly into perpetuity without gathering mold or reeking. At least that was my experience with several Kraft Singles I left in my refrigerator for more than two years as part of my trove of geriatric food. When I finally took them out of the fridge, the squares had shrunken in their plastic wrappers to about two thirds of their original size. They were hard and rigid, and resembled the bright orange plastic on a race track our kids play with. Some of the slices had cracked in half or lost a corner, and the surface was dotted with small beads of oil and strange clear crystals that looked sort of like salt, but weren't salty. I know because, reluctantly, I tasted them.

I wasn't sure what to make of my misshapen cheese slices. They brought to mind the disgruntled Wisconsin cheese dealers' insistence that Kraft's radical new product was "embalmed." Why did it shrink and what were those crystals? It was hard to explain what had happened to my cheese on the phone, so I took my petrified slices—which by now had reached the ripe old age of four years—to a cheese conference in Milwaukee and showed them to Steve Hill, director of cheese research and development at Kraft Foods, the company that had brought them to life—or not to life, as the case may be. A tall, blond food scientist who reminded me of actor Jeff Daniels, Hill's first reaction was to laugh affably at the notion that anyone would bother to keep cheese for that long, much less travel with it. He explained that the slices had probably shrunk due to moisture leaking out and proteins binding together, a reaction that also accounted for its plastic appearance. The clear crystals, he said, were lactose, a milk sugar which has a tendency to crystallize even in natural cheese.

As for the four-year absence of mold or a foul smell, Hill wasn't at all surprised about that. "Sorbic acid is a very good preservative," he said matter-of-factly, referring to one of the nondairy ingredients in processed cheese.

Sorbic acid is a chemical that was identified for its powerful antimicrobial properties in the forties; Kraft started adding sorbic acid to its cheese slices in the late sixties. It turned out that J. L. Kraft's 1916 assurance that his tins were "permanently keeping" had a caveat. They would remain unchanged only as long as you didn't open the package. Once you did, the clock started ticking, and the cheese could eventually become fodder for microorganisms. Sorbic acid became an insurance policy that this would never happen, that processed-cheese slices would always be 100 percent convenient and safe. It also guaranteed that the cheese's connection to the world of living things, of which food has always been a part, would be tenuous at best.

Harvey Wiley never penned any thoughts about processed

cheese, but he would have likely found the product lacking in "vital spark," a term he used to mean all the things that give food its power to nourish. Although Wiley understood a great deal about nutrition, he realized that food was still much more than he and other scientists understood it to be. Like life itself, there was something sacred and intangible about our daily sustenance, a reminder of our interconnectedness to the natural world and that we can't live without healthy soils and sunlight. As he wrote in a 1917 medical journal, "Food must not be dead. It must have a soul."

Of all the aging food items lingering, still and corpse-like, in my office and in the fridge, none seemed more lifeless than processed cheese.

4

—

Extruded and Gun Puffed

*Then we prepare our cereals in a method similar to what you do
in your own kitchen—just on a larger scale.*

—The Kellogg Company

On a picture-perfect early morning in June, thousands of
hungry people filed up Michigan Avenue in Battle Creek,
Michigan, making their way toward a ribbon of tables lined up
under small, white-topped tents. As a blinding sun rose, crowds
gathered two and three deep around neatly assembled rows of free
breakfast cereal portioned into bowls. A collection of people stood
behind each table, filling bowls. Some of them looked frantic, as
if they already knew what kind of pace was going to be required
to keep a steady supply of Cocoa Pebbles, Corn Pops, Raisin Bran,
and Corn Flakes coming. Just outside each tent, someone stood
ready with an opened carton of low-fat milk.

At the far end of the block, a troupe gyrated to dancehall reggae on a covered stage. Below them, a lively swaying crowd gathered on the street, marking the first time I witnessed anyone attempting to dance while eating a bowl of Frosted Flakes. Every so often, someone appeared with a large, bulging bag. "Pop-Tart?" they offered. Over a loudspeaker, barely drowning out the music, someone shouted, "Wake up, America. Battle Creek is serving breakfast!"

Or Battle Creek's version of it, anyway. The town's annual Cereal Festival—also known as "The World's Longest Breakfast Table"—is hosted by the three cereal companies with a presence in Battle Creek. Every year, Kellogg's, which has its global headquarters there, Post Foods, and Ralston (a store-brand manufacturer) send reps to hand out free cereal and greet the crowd. They donate thousands of boxes of some of their more popular varieties, much of it coming straight from their Battle Creek factories, making Cereal Fest a sort of celebration of local food. Industrial local food.

Although I'd arrived on Michigan Avenue at the festival's 8:00 AM scheduled start time, some people were already into their second bowl. Kevin, a retired Kellogg's plant employee, had had some Raisin Bran and then went in search of his primary target—Fruity Pebbles. "I was worried when I didn't see them at the first few tables, so I had to keep looking," he grinned. He wasn't the only grown man I saw enjoying brightly colored cereal. Scott, a middle-aged dad with long blond hair, a mustache, and a black Harley Davidson T-shirt, explained that he doesn't normally eat Froot Loops for breakfast. "This is what I grew up on, so it's like being a kid again," he said, diving into a half-empty bowl.

Many of the actual kids nearby were hovering around people dressed in plush cereal character suits, waiting to get their picture taken. The biggest masses of kids crowded around Tony the Tiger,

Toucan Sam, and Post's Sugar Bear. Considerably less in demand were the Keebler elf, Corn Flakes's Cornelius Rooster, and the mustached Pringles guy. The kids were briefly interrupted in their quest for photos when the dozens of Kellogg employees in attendance gathered with their characters for a massive group snapshot.

If Battle Creek residents are like most Americans, on any given day that isn't Cereal Fest, one fifth of them (or one third, if they're kids) will eat some type of breakfast cereal, a habit that helps fuel a $10 billion annual U.S. business and gives cereal its still-dominant place at the American breakfast table as well as an outsized share of space in the grocery cart. Items from the cereal aisle are the eighth most popular supermarket product—after soda, milk, bread, salty snacks, beer, wine, and cheese.

But soda, milk, and bread have nothing on cereal when it comes to sheer numbers of choices: there are more cereal iterations in the supermarket than any other product. My local King Soopers, owned by Kroger, the country's biggest grocery chain after Walmart, displays 215 different boxes, spanning both the regular and "natural" sections of the store. By contrast, the store sells 120 varieties of soda, 94 different sliced breads, 128 kinds of crackers, and a mere 34 types of peanut butter. Our landscape of choices in the supermarket is not an undisturbed democracy. There's no point in deliberating too long over what kind of oranges or tomatoes to buy. The decision has all but been made for us: navel or Valencia; vine-ripened, cherry, or Roma. In the cereal aisle, though, we live in the land of boundless opportunity.

Health Food

Considering the vast cornucopia of things that snap, crackle, and pop, it's hard to imagine a time when Americans didn't participate in the morning ritual of pouring cereal into a bowl. But like most processed food, boxed breakfast cereal is a twentieth-century creation. And it's one that's still uniquely relevant to America and its English-speaking allies. The United States, Canada, the United Kingdom, and Australia account for 6 percent of the world's population but more than half the world's breakfast cereal consumption, according to Cereal Partners Worldwide, a global venture between General Mills and Nestlé that's working on increasing consumption outside the English-speaking world.

Before the early 1900s, people did in fact consume quite a lot of cereal—just not the sort that could be eaten straight from the box. It came in the form of grains (typically whole) that had to be combined with water or milk and then cooked slowly. In the South, it was corn grits. In northern states, many Scottish and Irish immigrants didn't get through a day without oatmeal porridge. Around the world, many cultures still maintain these cereal traditions, eating *kasha* from wheat, barley, or oats in Eastern Europe; watery rice gruel called *congee* in China; *halim,* a sweetened wheat porridge flavored with cinnamon, in Iran and Turkey; and *pap,* a corn gruel used in a variety of southern African meals.

In addition to simple, cooked grains, in the nineteenth century, Americans, especially the well-to-do, ate a variety of things for breakfast that you aren't likely ever to see on a Denny's menu— wild pigeons, oysters, and stewed veal, for instance. They also feasted on considerable amounts of eggs, bacon, sausages, and fried ham. It was this voracious morning consumption of animal products that inspired a Battle Creek doctor and surgeon named John Harvey Kellogg to create the first modern, precooked, ready-to-eat breakfast cereal. A devout Seventh Day Adventist and vegetarian since the age of fourteen, Kellogg ran the Battle Creek Sanitarium, housed in an Italian Renaissance Revival building that still looms over downtown. The San, as it was called, operated as a hospital, spa, country club, and religious revival camp all at once. It attracted the elite of society, including President Warren Harding, Amelia Earhart, Henry Ford, Mary Todd Lincoln, and George Bernard Shaw.

Though he measured only 5 foot 4 inches tall, the cherubic doctor cultivated a larger-than-life presence, and the San's patients were drawn as much to Kellogg's outsized personality as his many novel ideas about health. An early adopter of preventive medicine, Kellogg espoused the benefits of sunlight and exercise, notions that were ahead of their time. Some of his other ideas, though, were flat-out bizarre, if not downright wrong. Kellogg believed, for instance, that vibrating chairs could improve circulation and cure constipation, and that coffee "cripples the liver." A Seventh Day Adventist, he argued against both sex and masturbation, believing that the "solitary vice" in particular led to everything from cancer of the womb and urinary diseases to impotence and epilepsy. He boasted that he and his wife, Ella, never consummated their forty-year marriage. Instead of conceiving their own children, the Kelloggs reared forty-two orphans.

The good doctor also had many thoughts about food, most of them considerably less eccentric. He believed the excessive consumption of animal products, especially for breakfast, was causing an epidemic of upset stomachs and other digestion-related maladies among those who could afford to eat an abundance of stewed veal and bacon. He was determined to find a healthier, low-fat, high-fiber way for his patients to start the day. Already serving oatmeal and whole-wheat porridges in his sanitarium, Kellogg wanted something new and exciting, a breakfast people could continue eating once they left the San. Ideally, it would be something that could be sent by mail and didn't require cooking.

John Harvey and Ella spent many hours in the San's kitchen, experimenting with new breakfast creations. One of their first products was a biscuit made of whole-wheat flour, oats, and cornmeal that had been mixed together into a dough and baked. Called granula, and later renamed granola, it was coarse, brittle, and nearly impossible to chew. To improve the texture, the couple strove to get the grain pieces much flatter—into a sort of flake shape—but this proved to be a challenge.

Just as it had for J. L. Kraft and his cheese, Kellogg's breakthrough came by accident. One day Ella boiled a batch of wheat kernels and then forgot to drain the water, leaving it soaking overnight. When Kellogg awoke he found that the wheat was quite soft and mushy. He ran it through a roller, and to his delight, each kernel of wheat stuck to the roller as a flattened flake. He scraped them off and baked them on a flat pan. The result was the first flaked cereal. The flakes were crispy and still somewhat leathery, but much less so than his earlier granola pucks. In 1897, Kellogg started the Sanitas Food Company to sell the cereal by mail order to former San patients who craved foods adhering to what Kellogg called "biologic eating."

It probably would have remained a small, quiet enterprise had Kellogg not hired his younger brother to work with him at the San. In contrast to the exceptionally charismatic and gregarious John Harvey, Will Keith Kellogg was taciturn and reclusive, with round glasses and an austere bearing. But he was infinitely more practical and business-savvy than his whimsical older brother. After years of fifteen-hour days, meager wages, and living at his brother's beck and call (reportedly suffering the indignity of taking dictation while John Harvey was on the toilet), Will Keith began asserting himself. The Sanitas Food Company was now selling corn flakes as a companion to the wheat variety, and while John Harvey was away on a trip in Europe, Will Keith seditiously added a touch of sugar to the recipe. When he found out, John Harvey was apoplectic. He demanded the sinful ingredient be removed, but W. K., overwhelmed by positive customer feedback, wisely ignored this request. The younger Kellogg saw enormous potential in Sanitas Toasted Corn Flakes and in 1906 bought out his brother's share of the company and founded what would eventually become the Kellogg Company, today's $13 billion breakfast and snack empire.

The brothers never reconciled. At one point, they sued each other over the use of the family name. W. K. ultimately prevailed. John Harvey continued to run the San, but made the ill-fated decision to finance an ambitious expansion, incurring large amounts of debt just before the stock market crash of 1929. When patients stopped showing up, he couldn't pay his bills, and in 1942 he was forced to sell the San building to the federal government, which today runs it as an office of the General Services Administration. He moved to Miami for a number of years, and in the final years of his life devolved into an even more eccentric character than he already was, regularly exercising outside wearing nothing but white cotton underwear. Several years before he died, he wrote a heartfelt letter to his brother apologizing for all the squabbles during their time together at the San. But W. K. didn't learn of the missive until too late. By this time he was blind from glaucoma, and his associates hadn't told him about the letter, thinking it would upset him. He became aware of it, to his regret, only after John Harvey died in his sleep in 1943 at the age of ninety-one.

Although John Harvey's presence loomed over Battle Creek in its early days, it was Will Keith who created the enduring legacy. Today, the Kellogg name appearing on a Battle Creek auditorium, middle school, community college, airport, and large foundation building all refer to the younger sibling. When W. K. died in 1951, also at the age of ninety-one, the Kellogg Company laid his body in an open mahogany casket for a day in the main lobby of its Battle Creek headquarters.

Machine Food

Walk down a cereal aisle today or go onto a brand's Web site, and you will quickly learn that breakfast cereal is one of the healthiest ways to start the day, chock full of nutrients and containing minimal fat. "Made with wholesome grains," says Kellogg's on its Web site. "Kellogg's cereals help your family start the morning with energy by delivering a number of vital, take-on-the-day nutrients—nutrients that many of us, especially children, otherwise might miss." It sounds fantastic. But what you don't often hear is that most of these "take-on-the-day" nutrients are synthetic versions added to the product, often sprayed on after processing. It's nearly impossible to find a box of cereal in the supermarket that doesn't have an alphabet soup of manufactured vitamins and minerals, unless you're in the natural section, where about half the boxes are fortified.

But if cereals are so nourishing, why do they need so much help? The answer can be found in cereal factories, where the making of this food has evolved dramatically from the early days in Battle Creek.

Compared to today's breakfast cereals, the Kellogg brothers' early wheat and corn flakes were crudely made, nutritious creations. They were crafted from whole grains, little if any sugar, and only a few other ingredients. They packed a lot of fiber, as John Harvey had intended, and were likely to boast naturally occurring B vitamins and, in the case of Corn Flakes, vitamin C, though no one at the time knew how to measure such things.

They wouldn't stay like this for long, though.

The emerging system of warehousing and centralized production dictated that packaged products like breakfast cereal be able to survive for several months. On this count, Corn Flakes didn't fare so well. After just a month, boxes could develop a rancid odor due to the presence of oil from the germ portion of the corn. If Corn Flakes lovers ate their cereal within several weeks of it being produced, this wasn't a problem, but W. K. knew he needed a product with better shelf life.

In 1905, he changed the Corn Flakes recipe in a critical way, eliminating the problematic corn germ, as well as the bran. He used only the starchy center, what he referred to as "the sweetheart of the corn," personified on boxes by a farm girl clutching a freshly picked sheaf. This served to lengthen significantly the amount of time Corn Flakes could sit in warehouses or on grocers' shelves but compromised the vitamins housed in the germ and the fiber residing in the bran. Since vitamins and fiber were yet to be identified in food, it's likely W. K. didn't realize he was making a nutritional trade-off. For all he knew, his customers were getting a much-improved product.

Food scientists have since figured out ways to use heat and preservatives to deactivate the enzymes that cause corn and wheat oil to go rancid, though Kellogg's never did return to using whole corn. W. K.'s fix for the cereal's shelf-life problem stands as one of the earliest examples of what would become a central paradox of the food processing industry—the fact that nutrition and convenience are sometimes deeply at odds with one another.

As for the crudely processed part, that didn't last long either. The first transformative technology was something called gun puffing. In 1904 the Quaker Oats Company introduced new machinery at the World's Fair in St. Louis. It hauled in a row of bronze army-surplus cannons, filled them with white rice, and then applied heat. As the cannons got hot, large amounts of pressure built up inside their chambers. When the chambers opened, the resulting rapid drop in pressure forced the rice to explode, letting off a loud boom and launching a storm of airy puffs over the crowd—Puffed Rice, the company's first ready-to-eat breakfast cereal. A Quaker poster at the fair pronounced the new technology "The Eighth Wonder of the World."

Despite the awe-inspiring display, the exploded cereal failed to catch on with consumers and languished in the marketplace for nearly a decade—that is, until 1913, when the company began an ad campaign featuring the tagline "The Grains That Are Shot from Guns." Like Kraft's processed-cheese ads, the campaign leveraged a growing sense of wonderment toward technological innovation. Quaker ran ads highlighting the scientific novelty of Puffed Rice, sometimes featuring the lab-coated former University of Minnesota chemical biologist who had created the product. On boxes, customers were informed that their cereal was "steam exploded" to "8 times normal size." Sales increased tenfold in five years, placing both Puffed Rice and Puffed Wheat in the breakfast mainstream.

Quaker's "Guns" campaign reflected a broad sense of pride in the food industry about the ways science was being used to create what were billed as improved food products. In 1919, canned-food manufacturers declared their products "The Miracle at Your Table." In one ad, six housewives gazed in admiration as a bespectacled scientist at the National Canners Association's laboratory stared intently into a beaker. For several years, starting in 1905, boxes of C. W. Post's Grape-Nuts bore this invitation: FACTORY ALWAYS OPEN TO VISITORS.

Some hundred years later, none of the cereal companies I contacted for this book, including Post Holdings, the maker of Grape-

Nuts, were able to offer tours of their factories, though several older residents of Battle Creek fondly told me about excursions through the Kellogg plant in the fifties and sixties.

Americans today know far less about how our breakfasts are made than did our less educated ancestors living in far less techno-savvy times. In 1913, most Americans knew more or less what a gun-puffing machine was, thanks to Quaker. Today, almost nobody does. Mention automatic single-shot guns, automatic multiple-shot, and continuous guns—the types of gun-puffing machinery in use today—and people are likely to think you're planning a Civil War reenactment, not referring to the production of Sugar Smacks or Cheerios.

Rather than tout their technological prowess, modern food companies seek to highlight the pastoral origins of their products. Images of amber waves of wheat, lush fruit, and happy farm animals beckon from packages, and ads tout "natural goodness." On the subject of what happens between the picturesque farms and the supermarket aisles, manufacturers are generally mum.

Customers of Bear River Valley all-natural cereals, for instance, could easily assume that these products are produced in a careful operation in a tiny stream-studded town in northern Utah called Bear River City, population 853. Product packages feature images of mountains and rivers. On the Web site, a map shows the exact location of this bucolic outpost. Bear River City is a real town, after all. It's just that Bear Valley all-natural cereals aren't produced there. They're made eight miles north in Tremonton, in a colossal $100 million cereal plant owned by MOM Brands (formerly Malt-O-Meal). It's a site of production for most of the company's other cereals as well—brands like Apple Zings, Muffin Tops, Coco Roos, and Honey Buzzers.

After gun puffers came extrusion machines, which were first

created in the thirties and more widely adopted as a way to cook
and process cereal in the sixties. This breakthrough made everything before it seem antediluvian. Extrusion machines could take
multiple ingredients, mix them together rapidly and form cereal
into a potentially endless array of fun-filled shapes. They ushered
in an era of breakfast shaped as diamonds, clovers, and letters of
the alphabet, helping to capture the attention of children. The
real beauty of these machines, though, was that they slashed costs.
Instead of the seven hours it took to cook and process grains, the
whole operation could be done in one continuous, twenty-minute
step (today, raw materials zip through extruders in more like fifteen to sixty seconds). This meant not only a higher output, but
fewer machines, less square footage, and reduced energy usage,
resulting in lower production costs and greater profitability. As
an extra bonus, extruders served to kill bacteria, maximizing food
safety.

Some cereals, like those MOM Brands Apple Zings and Honey
Buzzers, and many lower-priced and store brands, are created
almost entirely within the tubular confines of extrusion machines.
Ingredients flow into one end, and an endless succession of crisp,
puffed shapes emerge—extrude—from a die at the other. Other
varieties are partially cooked into pellets in extruders and then
further processed—either gun puffed (Cheerios, Froot Loops, and
Cascadian Farms Fruitful O's) or flattened and toasted (some raisin brans).

Extrusion is undoubtedly the harshest and most nutritionally
devastating way to process cereal. These are not gentle machines.
They look a bit like oversized jackhammers turned on their sides.
Inside the long, steel barrel, starch, sugar, and protein molecules
are ripped apart by twisting screws that generate large amounts
of heat and pressure. Think of extrusion as a molecular melting

pot. One food scientist likened starch molecules in an extruder to exploding water balloons. "You swell the starch granule and then it breaks, spilling its guts into the solution it's in," he told me. Such damage is quite intentional, since it's what allows ingredients to meld together quickly, forming a thick, homogenized mass. The process is often referred to as "plasticization"—which neatly sums up the nutritional gist of what happens inside an extruder.

According to a 2009 study done by Mian Riaz, a food science professor at Texas A&M and one of the country's leading extrusion experts, the nutrients most vulnerable to extrusion, which is also used for puffed snacks such as Cheetos and snack bars, are vitamins A, B1 (or thiamine), C, E, and folate, the natural form of folic acid. The diminishment of vitamin B1 is particularly unfortunate because cereal grains are one of our most important sources of this nutrient. The degree of damage varies widely but can sometimes reach 100 percent. Naturally occurring fiber and phytochemicals, such as the antioxidants present in oats, also fare poorly.

Beyond the extruder, steam cooking under pressure, drying, and toasting, which occurs at a fiery 525°F to 625°F, subject grains and other raw materials to much higher temperatures than anyone normally would at home. Home cooks generally don't need to evaporate every last drop of water from food in order to give it a long shelf life. Most commercial cereal has been dried so thoroughly that it is virtually immune to decomposition—which explains why the contents of the six-year-old boxes sitting in my office look exactly the same as the day I bought them.

The assault on cereal nutrition doesn't end with processing: surviving vitamins have to endure the box. Breakfast cereal can sit around for up to nine months before anyone buys it, and over time most vitamins naturally degrade and lose their potency. For insurance, manufacturers compensate by adding in synthetic vitamins at overages up to two and a half times what's listed on the package. In other words, if the label says you'll get 30 percent of the recommended daily allowance (RDA) of vitamin C, there might have been 75 percent RDA added, just to ensure that by the time you spoon the crunchy squares into your mouth, 30 percent is there. Minerals like iron, calcium, and zinc aren't as delicate as vitamins, so they don't need to be added in at such high levels.

All this nutrient loss is the collateral damage of high-output industrial production. So, too, with naturally occurring flavor. What little flavor develops in a sixty-second extrusion-cooking process tends to "flash off," along with the superhot moisture in the dough. And so, the only foolproof way to ensure that cereal doesn't taste like the box it's sold in is to use more durable, manufactured flavorings. The same goes for colorings. Pellets and cereal bits sometimes emerge from extruders appearing gray or unappealingly dull, which accounts for the Yellow #5 used in Kellogg's Smart Start cereal, even though the product doesn't look very yellow.

Soft and Smooth

In National Lampoon's *Vegas Vacation,* Clark Griswold, Hollywood's only leading food scientist character, finances a week of temptation and misadventure in Sin City with a generous bonus he received from inventing the "Crunch Enhancer." A game-changing "nonnutritive cereal varnish," it ensures that flakes, squares, and O's won't ever get soggy. "It's semipermeable," Clark tells a co-worker. "It's not osmotic. What it does is it coats and seals the flake, prevents the milk from penetrating it."

Real food scientists would be envious of Clark. They have yet to arrive at any such foolproof solution for what they call short "bowl life." All but the most exceptionally fast eaters know that once in contact with milk or some other liquid, cereal soon dissolves into a pile of soggy mush. This is especially true for varieties that have gone through an extruder. The damaged starches simply fall apart, leaving behind a bowl of glop. Manufacturers do their best to lengthen the amount of time that can pass before this happens: they toast at high temperatures and spray the cereal with a fine coating of sugar to serve as a moisture barrier. But the ultimate "crunch enhancer" remains elusive.

While the idea of soft, predigested food may sound like a good thing—the work of breaking it down has already been done for us!—this turns out not to be the case. The human gastrointestinal tract has spent tens of thousands of years digesting crunchy, fibrous foods, and along the way it's come to appreciate the challenge. Depriving our stomach of its gastric duties by giving it disassembled food appears profoundly to alter energy metabolism and the dynamics of hunger and satiety. When starches arrive in our stomachs already broken down, they enter our bloodstream rapidly, causing a spike in insulin and potentially fostering a dynamic that can lead to the condition known as insulin resistance, which is a precursor to type II diabetes.

It's not as simple as saying that hyperprocessed starchy foods are bad for us. A little bit of damage to the starch is actually a good thing. Humans can't effectively digest uncooked grains. No one eats raw corn on the cob or gnaws on wheat kernels. Crushing, flour milling, and normal cooking all start the breakdown process in a helpful way, allowing our bodies to access a grain's nutrients. But industrial processes like extrusion and gun puffing are extreme; they dismantle foods to the point where there's not much left for our digestive systems to do.

Since 8 percent of the U.S. population suffers from type II diabetes and 35 percent of us have metabolic syndrome, a sort of pre-diabetes, the issue isn't simply academic. It's also thought there's a link between intensive food processing and weight gain, since one of the ways we burn calories is through the process of diges-tion. Anywhere between 5 and 15 percent of a person's daily calo-rie expenditure comes simply from eating. Hyperprocessed food requires considerably fewer calories to assimilate.

Ample research supports this notion, which runs counter to the theory that a calorie is just a calorie, regardless of whether it comes from orange slices or a bottle of orange Gatorade. For twenty-two weeks, a team of Japanese researchers fed two groups of young rats a diet of rat chow that was nutritionally identical, except that one ration contained the standard hard-to-chew pellets and the other contained pellets that had been puffed up with air, making them softer and requiring about half as much force to chew. If all calo-ries are equal, then both groups should have grown at roughly the same rate and to the same eventual size. But that didn't happen. The rats that consumed the airy pellets gained more weight than their brethren. The weight gain was gradual at first, but by the end of the experiment the rats eating soft pellets weighed about 6 per-cent more and had 30 percent more abdominal fat, enough to be classified as obese. The reason for this appeared to be that the soft pellets induced lower postmeal metabolic rates and diminished levels of what's known as thermogenesis, the production of heat by our bodies.

A California study that evaluated human responses to a "whole" meal versus a "processed" one yielded similar findings. Eighteen people were given two different meals consisting of the same num-ber of calories on different days. The "whole" meal consisted of cheddar cheese and "100 percent natural" multigrain bread made with stone-ground whole-wheat flour (which is generally coarser than standard-ground flour), sunflower seeds, oats, barley, corn, and millet. The "processed" meal was white bread and slices of American cheese. Although the participants said they found both meals equally satiating, their metabolic rates after the two meals were quite different—almost 50 percent higher for the whole-food meal. If you extrapolate that over many meals and many months, it's not hard to see how losing as much as half of that 5 to 15 percent digestion-related calorie burning could lead to slow, creeping weight gain, the sort you don't even realize is happening because it is related entirely to the kind of food you eat, not the quantity.

Empty Calories

So how do you find out how much, if any, of your cereal's vitamins come from Mother Nature? Good luck. The nutrition label on the side panel will be of no help; labeling rules don't require manufacturers to distinguish the source of the nutrient. All you can do is wonder.

Which is exactly what Robert Choate did. Before nutrition information of any sort was required on packaging, Choate, a senior staff member in the Nixon White House, rang the alarm about the deluge of nutritionally suspect, sweetened cereals that had flooded the market during the previous fifteen years, thanks to new technology for sugar coating. In the fifties, General Mills introduced Jets, Frosty O's, and Cocoa Puffs; Kellogg's came out with Corn Pops, Frosted Flakes, Sugar Smacks, and Cocoa Krispies, following it up in the sixties with Froot Loops, Apple Jacks, and Frosted Sugar Stars. "I watch the TV commercials on Saturday morning and get really mad," Choate said in 1970. "The industry brainwashes children into demanding the least worthwhile products . . . Nutrition doesn't snap crackle, or pop."

Handsome, with a deep voice and dark mane of slicked hair, Choate was born into a blueblood New England family (a relative founded the Choate Rosemary Hall prep school in Connecticut). He attended Phillips Exeter Academy prep school in New Hampshire and served in the Navy during World War II. Following that, he earned a civil engineering degree from the University of California at Berkeley and started a construction business in Phoenix, earning a small fortune in real estate and adding to the windfall he'd inherited from his father, a successful Boston newspaper publisher. Choate's life took an unexpected turn when he contracted hepatitis in the late fifties and spent a year recovering from it. While bedridden, he read the autobiography of Walter White, an early leader of the National Association for the Advancement of Colored People (NAACP) who had African ancestry but looked Caucasian, with blond hair and blue eyes. This and other books about racial justice inspired Choate to reevaluate his life. He decided to pursue a new career working to alleviate poverty. As part of that, he initiated a national study on malnutrition, which sparked his interest in food and helped land him in the Nixon White House.

Taking note of the central position that boxed cereal had assumed at the American breakfast table, Choate hired a lab to calculate the amounts of nine important nutrients contained in sixty different breakfast cereals. On July 23, 1970, he presented his findings to the Senate Commerce Committee chaired by Senator Frank Moss of Utah. The results were not encouraging for cereal companies: On a scale from 0 to 700, with 700 being the most nutritious, two thirds of the tested products ranked below 100, including popular varieties such as Corn Flakes, Rice Krispies, Sugar Frosted Flakes, Cheerios, Wheaties, and Shredded Wheat. Many cereals, Choate told the committee, "fatten but do little to prevent malnutrition." They were, he said, "empty calories."

Choate's testimony was widely reported in the media. Cereal makers were both embarrassed and enraged, but they lacked solid counterarguments with which to defend themselves. Kellogg's led an attack on Choate's credentials but never disputed the specifics of his findings. "Civil engineer Choate's theories and so-called formula might be meaningful for digging a mine shaft," the company's director of research told the Associated Press. "But they are completely valueless as a yardstick for measuring the nutritional value of any type of food." Other companies charged that Choate had failed to take into account the nutrients from milk, cereal's trusted partner. Choate had also neglected to compare the nutritional value of cereal, they said, with other breakfast foods, such as donuts.

This marked the first use of the better-than-a-donut defense, though it wouldn't be the last. The argument resurfaced four decades later in a doomed attempt to explain why Froot Loops is a healthy product. In 2009, the food industry created a nutrition ratings program called Smart Choices that was to give better-for-you products a green check mark on their packages. Froot Loops and Apple Jacks, it turned out, were among those products. In an interview with the *New York Times,* Eileen Kennedy, then the dean of Tufts University's nutrition school and head of the Smart Choices board, offered up this explanation: "You're rushing around [at the supermarket], you're trying to think about healthy eating for your kids and you have a choice between a doughnut and a cereal. So Froot Loops is a better choice." Several months later, the program, having made some un-smart choices, was history. The food industry quietly walked away from it.

General Foods, the maker of Post cereals, took a different tack in its response to Choate's 1970 Senate assault. The company disagreed with the notion that its sweetened cereals like Honeycomb, Crispy Critters, and Super Sugar Crisp, now named Golden Crisp, would lead to excess sugar consumption among children. "In our opinion, exactly the opposite is true," the company said at the time. "Presweetened cereals provide a measure of control over sugar intake that is not present when the young consumer sweetens his own."

General Foods's intimation that children might dump the entire contents of a sugar bowl into their morning meals failed to gain much traction in the public imagination. Choate's testimony, on the other hand, stirred outrage, and cereal manufacturers felt compelled to take action. They did this by adding vitamins and minerals to their formulas. Some cereals already included some, so manufacturers added more; others had none and welcomed fortification for the first time. The breakfast cereal aisle was transformed into the most heavily fortified real estate in the supermarket. Two years later, more than half of the forty cereals Choate had execrated had been fortified and returned to the shelves as new and improved.

But Choate wasn't finished. He hired two leading nutrition professors, one at the University of Nebraska and the other at the University of Georgia, to conduct a series of cereal feeding tests with juvenile rats. More bad news for cereal companies. Rats on a diet of all but six of forty cereals showed little or no growth, even when their food was supplemented with a separate vitamin and mineral mixture. Armed with these latest results, Choate returned to the Senate for another newsworthy appearance. At one point he poured the heaping amount of sugar contained in a box of Post Pebbles onto a table. He railed against prioritizing profits over the nation's health. "The temptation of the dollar is greater than the will to nourish the population," he warned.

A few years later, *Consumer Reports* sought to replicate Cho-ate's tests, putting groups of rats on forty-four brands of cereal, most of them fortified, for at least twelve weeks. The magazine reported that half the cereals proved nutritionally "deficient" in the experiments and the other half were "adequate," although rats eating some of this latter category still displayed minor symptoms of nutritional deficiency, such as graying or browning of the hair and nervousness. The authors acknowledged that it might seem strange to readers that a cereal like Product 19, which was fortified with 100 percent of the RDA for vitamins and minerals, appeared at the bottom of the rankings. But human and rat bodies require a complex web of nutrients beyond what had been added, they explained. The magazine referred to this web of then-unknown substances as "intrinsic factors." Scientists today would call them carotenoids, flavonols, and polyphenols, members of an enor-mous universe of beneficial plant chemicals, the likes of which companies are still trying to figure out how to incorporate into their food.

No one has since done these sorts of comprehensive feeding tests on breakfast cereal, or any other processed food for that matter. So we don't know how cereals would fare today. To some degree, nutrition in the category has improved since the seven-ties. Many cereals still list sugar as a primary ingredient, but there tends to be less of it per serving than there was three decades ago. All of General Mills's kids' cereals (Cookie Crisp, Trix, Boo Berry, etc.) now contain no more than 10 grams of sugar per serving and include some amount of whole grains, mirroring a broader indus-try effort to replace some of the stripped-down wheat and corn with more intact varieties. But whole grains are useful only if the good stuff in them—the vitamins, minerals, phytochemicals, and fiber—survives both processing and shelf life, and figuring out to what extent that has happened isn't easy.

When I inquired, Kellogg's, General Mills, and Quaker (owned by Pepsi) said they weren't interested in discussing the subject. However, Post Foods, the nation's third-largest cereal company, was willing to provide a chart showing the levels of naturally occurring B vitamins—those most abundant in wheat—in its Great Grains cereal. What it showed, not surprisingly, is that most of the vitamins you get by eating this cereal come from what's been added. According to the box, one serving of Great Grains gives you between 25 and 35 percent of the RDA for thiamine (B1), riboflavin (B2), niacin (B3), B6, and folic acid. Without synthetic additions, the numbers would be much lower—10 percent, 2 percent, 8 percent, 4 percent, and zero, respectively. And these levels represent what you'll get from a freshly made cereal, not one that's been hanging around for six months.

For most commercial cereals, this may be as good as it gets. If you can get beyond its relatively high sugar content, Great Grains stands out as one of the healthier offerings on the market, having been recently reformulated into, as Post puts it, "less processed nutrition you can see." Mark Izzo, Post's director of research and development, explained to me how the company now uses whole, intact wheat berries to make its flakes, not just an amalgamation of various forms of ground-up flour. "We get the wheat kernels in via rail car and then we batch cook them with a gentle steaming process, no different than what you could do at home with a pressure cooker. Then we roll them and bake them," he said. No extrusion and no gun puffing (both of which the company still uses for many of its other products, like Honeycomb and Alpha-Bits). The result is large quantities of whole grains—32 to 40 grams per serving, depending on the variety, and a lot more than what's in

Cheerios (14 grams), Kellogg's Raisin Bran (15 grams), and Special K (9 grams). Great Grains still contains the chemical preservative BHT. "We have very active programs working on replacing that," Izzo assured me.

Post learned the value of cereals made from gently processed whole grains several years ago when it did testing to see if there were naturally occurring antioxidants in its products. The cereals registering significant levels were those getting an antioxidant boost from raisins (which, of course, aren't cereal) and those that were minimally processed, at least relatively speaking—Shredded Wheat and Grape Nuts. Post did this testing in 2009, before the re-launch of Great Grains, so that brand wasn't tested.

Yet still, much like Great Grains, Shredded Wheat and Grape Nuts—products at the cereal aisle's nutritional pinnacle—offer far fewer intrinsic vitamins that what you'd find in an equivalent amount of whole-wheat flour. It's hardly a ringing endorsement for a category often touted as a bastion of wholesome and nutritious choices.

Dairy industry people are generally big fans of breakfast cereal; it helps boost milk consumption. During a presentation at IFT 12, held in Las Vegas, Nancy Auestad, a vice president at the Dairy Research Institute, highlighted breakfast cereal as a "critically important source of nutrients for the American population." And in a sense, it is. But the origin of the vitamin B listed on the side of your cereal box isn't necessarily an American-grown grain. When it comes to nutrients, cereal makers get large doses of assistance from faraway places, predominantly factories in China.

5

——

Putting Humpty Dumpty
Back Together Again

The active ingredient in broccoli is broccoli.

—David Katz, director of the Yale University
Prevention Research Center

Every day at the kinetic Port of Melbourne, Australia, a container vessel the length of your average strip mall sets sail. Inside steel boxes stacked high atop its deck are all sorts of things you'd expect to be exported out of Australia—natural materials like timber and pulp; agricultural commodities like barley and cotton; bottles of Australian wine; and that most enduring of Aussie exports, sheep wool. Sailing steadily, the ship makes its way northward, skirting Australia's eastern coast, winding through the islands of Papua New Guinea, and charting a course into the open waters of the Pacific. After two and a half weeks, it arrives within

China's nautical borders and lands at the port of Shanghai. There isn't a lot of historical precedent for wool being transported along this route. Until recently, most of what was shorn from the backs of Australia's 73 million sheep went to Europe and North America. Today most of it finds its way to China. Firms there buy between 70 percent and 80 percent of the wool Australia produces, up from less than 25 percent in 1999.

This is about more than the manufacture of sweaters. Much of the wool China buys is equally valued for the grease embedded in it as for the wool itself. As ducks secrete oil to make their feathers waterproof, sheep produce a similar fatty substance that helps protect them from harsh weather. Australia's wool is particularly greasy, and this grease—or various derivatives of it—is useful for making a whole slew of industrial and consumer products. Some portions go to produce lubricants for machinery and waterproofing for boats. Others, like lanolin, become lip gloss, moisturizer, and sunscreens.

And there's another end point for this grease—something hardly anyone would ever associate with wool. At a factory in Dongyang, a burgeoning industrial center on China's eastern coast, the grease's cholesterol component is used to make vitamin D. Zhejiang Garden Biochemical is the world's largest maker of this vitamin—one that goes into nearly all the milk Americans consume (including organic varieties), as well as many of our breakfast cereals, breads, bars, margarine, and other dairy products.

And so it is that our milk doesn't come entirely from cows. For some small part of it, we're indebted to the greasy backs of Australia's sheep.

Vital Amines

Vitamins represent one of the great scientific milestones of the twentieth century, dating back to the distinguished work of two men: Nobel Prize–winning Dutch physician Christiaan Eijkman and the Polish biochemist Casimir Funk. Eijkman made the first critical discovery just before the turn of the twentieth century in the Dutch East Indies, now Indonesia. Sent there by the Dutch government, he was charged with locating the cause of and ideally a cure for the neurological disease beriberi, which had become epidemic in the region. After several years of investigation, he realized that when chickens were fed the white, or polished, rice that had recently become more available thanks to the development of polishing machinery, they came down with the disease. But when their feed was switched to whole brown rice, the beriberi went away.

It was already generally known that food had the ability to cure disease, but nobody really understood why. English sailors recognized that the debilitating disease scurvy could be eradicated with citrus fruit, hence the term *limey* as a nickname for the British. People in Northern Europe learned to treat the skeletal deformities from rickets with cod liver oil and to reverse night blindness with liver. Eijkman correctly identified the cause of beriberi, but he misunderstood the principles of how the phenomenon worked, thinking it might be due to a substance in the husks of brown rice that disabled rice's poisonous effects.

Working in a lab in London some years later, Casimir Funk read a report on Eijkman's findings and decided to pick up where the Dutchman had left off. In 1911, he succeeded in identifying and isolating from whole rice the substance that Eijkman, who was a doctor and didn't possess the tools of chemistry, couldn't find. Funk called this substance a "vitamine," a fusion of the term *vita* (Latin for life) and *amine,* meaning nitrogen compound. Found naturally in whole rice, most of this "vitamine" was lost when rice was processed. Later it became known as "thio-vitamine," meaning sulfur-containing vitamin, and then thiamine and B1, its current appellations. Funk's groundbreaking work inspired a new field of research that over the course of the next thirty-five years succeeded in pinpointing the thirteen vitamins and some fourteen minerals we now consider essential.

The initial means for accessing these vitamins was to get them from the foods in which they were found abundantly. D and A came from cod liver oil, C from oranges, B1 from rice, and B2 from liver. But in the thirties, mostly in Europe, chemists figured out how to synthesize these compounds in a lab, bypassing food sources entirely. Companies like Roche in Switzerland and the German chemical firm BASF established what would eventually become hundred-million-dollar vitamin manufacturing businesses. Not long after, food companies began experimenting with these new-fangled health boosters. In 1938, Kellogg's Pep became one of the first breakfast cereals to be fortified with vitamins.

The new, high-tech additions also started showing up in hot dogs, puddings, and soda. In 1941, the Doughnut Corporation, which at the time had a near-monopoly on the sale of donuts in the United States (it would be another decade before the nation's first Dunkin Donuts opened its doors), ran an eye-catching ad campaign for "Vitamin Donuts." Fortified with ample amounts of Casimir Funk's thiamine and smaller amounts of B3 and iron, the donuts promised "pep and vigor" to a wartime population ("pep" being, presumably, a popular word at the time).

For the most part, though, these initial forays into fortification were dry runs. As the Canadian food historian Harvey Levenstein writes in his book *Paradox of Plenty,* many manufacturers remained hesitant about fortification, fearing it would amount to an acknowledgment "that their critics had been right, that processing often did deprive food of nutrients." When, at the urging of the government, flour millers started adding the B vitamins and iron to white flour in 1941, it seemed exactly this sort of concession—a confirmation of Harvey Wiley's conviction that white flour was a nutritional vacuum. For six decades, since the dawn of steel roller mills, milling companies had been casting off nearly all of wheat's nutrients to make the fine, silken flour people increasingly desired. No one could have proved it in the second half of the nineteenth century, but by the 1940s the subject of white flour's nutritional inferiority was no longer in dispute.

I came across a 1952 article on this subject in a journal called *Baking Industry*. It was written by Dr. W. B. Bradley, the scientific director for the American Institute of Baking, and in it he offered the sort of reflective commentary you don't usually find in industry trade journals. Bradley speculated that if knowledge of nutrition science had managed to arrive before the equipment for developing white flour, the baking industry's future might have been quite different. The industry, he said, "probably would have hesitated to set in operation" the production of "a less nutritious staff of life."

The article revealed Dr. Bradley to be an optimist. By his logic, J. L. Kraft would have remained a mere cheese distributor had he known his processed product was going to be less beneficial than natural varieties. And W. K. Kellogg would have found another way to increase Corn Flakes's shelf life or simply been content with a short-lived product. We have no way of knowing whether food-industry pioneers would have decided against uncorking the genie bottle were they to have known the contents. We do know that the food industry's exponential growth throughout the twenti- eth century was driven not by a quest for healthier products, even when more advanced knowledge of nutrition was available, but by a reach for elevated levels of innovation and creativity that would ultimately lead to greater sales and profits.

In any case, it would be quite some time before the government attempted to break up America's long-standing love affair with white bread. It wasn't until 2005 that the U.S. Department of Agriculture's (USDA's) Dietary Guidelines made the first serious bid to emphasize the superiority of unrefined, whole grains—an effort that's been modestly successful. Our consumption of whole grains is up 20 percent since 2005, though it still accounts for only 13 percent of the total grains we eat. (Yet this is an improvement over the barely registerable levels of whole grains we consumed in 1950.)

Made in China

Vitamins and minerals are now a celebrated addition to many food products. Companies need not declare why the nutrients are in there; often just the simple fact of their presence is enough. "Marshmallow Pebbles is a wholesome, sweetened rice cereal," Post declares on its Web site. "It is low in fat, gluten free, cholesterol free, and provides 10 essential vitamins and minerals. It is also an 'Excellent Source of Vitamin D'!" Kraft's MilkBite Bars boast that they have "the calcium of an 8 oz glass of milk," without clarifying that some of the calcium comes from calcium phosphate and calcium caseinate, not entirely from the milk in the product, which makes an appearance on the ingredient list in the form of cream and skim milk.

As one glance at those World War II Vitamin Donut posters tells you, this is an old trick but one more prevalent than ever. That the marketability of added nutrients is at an all-time high is evidenced by the fact that artificial sweeteners now contain them. Packets of Splenda Essentials have B1, B5, and B6 "to help support a healthy metabolism." In fact, vitamins and minerals are so thoroughly embraced that they're the only synthetic ingredients with carte blanche approval for inclusion in certified organic products, even when those vitamins and minerals are produced with genetically modified (GM) bacteria or have been synthesized from noxious petrochemicals. GM technology and toxic chemicals are otherwise banned from organics.

Vitamins and minerals form such ubiquitous background music in the supermarket that we no longer think about how they got there or whether they're actually the same thing as the food substances they were modeled on. Even people you'd assume would know something about how vitamins come to be aren't necessarily aware. At a conference called SupplySide West in Las Vegas, I asked the retired founder of a company selling specialized mixes of vitamins how his products are made. "I don't know and I don't care," he bellowed, cradling a late afternoon beer and seemingly not in the mood for serious questions. Some months later, when I e-mailed Marius Cuming, media manager at Australian Wool Innovation, an organization of the country's wool growers, to ask him about wool's path through China and into our vitamins, he responded that he was pretty sure wool didn't have anything to do with vitamin D. I forwarded him some info and he replied, "The experienced wool scientists at AWI are certainly raising eyebrows. . . . I didn't know that D3 was manufactured from wool lanolin— what a great health claim!"

Cuming understood intuitively: health claims are exactly what make vitamins a $3 billion worldwide business.

Contrary to popular assumption, vitamins added to products in the manufacturing process almost never come from the foods that contain them, or any other foods for that matter. Just as vitamin D doesn't begin with egg yolks or cod liver oil, vitamin C has not been squeezed from an orange, and vitamin A has absolutely nothing to do with a carrot. Nor does calcium somehow originate from milk or "old crushed cow bones," as one friend suggested. Vitamins can be derived from foods, but doing so is inefficient and wildly expensive, so hardly anyone does it. A French company called Naturex sells vitamin C from acerola cherries, a fruit loaded with the nutrient. But their product ends up being five times more expensive than the standard synthetic variety. Getting manufactured vitamins from food might also be problematic from a supply-demand point of view, since food-based vitamins could potentially gobble up edible crops. But there are ways around this, such as with acerola cherries, which are eaten in the subtropical countries they're native to but are not consumed in the United States.

Today, the short answer to the question of where vitamins come from is China. Chinese firms account for roughly half of all global vitamin production, according to Leatherhead Food Research in London, though for some specific varieties it's much higher than that. Half a dozen Chinese companies, for instance, are making 90 percent of the vitamin C, or ascorbic acid, that goes into food as both a vitamin and a preservative. Along with Zhejiang Garden's dominance in vitamin D, Chinese firms make the bulk of the world's B1, B2, and B12. The European companies also have jumbo-sized Chinese facilities. Lonza, a Swiss company responsible for manufacturing the majority of the world's vitamin B3, does so at a plant in Guangzhou.

Such geographic concentration may not be all that surprising when you consider the production of food additives in general. Chinese firms manufacture 60 percent of the world's xanthan gum, two thirds of its monosodium glutamate (MSG), 25 percent of starches, and 40 percent of emulsifiers, stabilizers, and thickeners, according to Leatherhead. And they've done it without incident. Although China has been implicated in one food contamination scandal after another, none of them has ever involved vitamins or other U.S.-bound food additives (though in 2007, some of America's pets ate wheat flour laced with melamine). In fact, the success of China's global food ingredient business and its near-takeover of certain markets has gone without much attention of any sort.

The last vitamin C plant operating in the United States—which was owned by Roche and made ascorbic acid by brewing different chemicals together—stopped production in 2005. It was located in the leafy, suburban hamlet of Belvidere, New Jersey. In 1999, the Environmental Protection Agency (EPA) reported that the plant, which was situated a mile from the local high school, was emitting large amounts of hazardous air pollutants—methanol, chloroform, and toluene—into the surrounding air. Both the EPA and State of New Jersey issued fines. Then a sweeping federal vitamin price-fixing suit, of which Roche was a part, made the vitamin business even more unattractive. In 2003, Roche unloaded its operations to the Dutch conglomerate Royal DSM, which switched the Belvidere plant to the less-polluting chore of mixing blends of vitamins. The change created a void in the market that Chinese companies quickly stepped up to fill.

The only remaining vitamin plant of any sort in the United States is a DSM beta-carotene facility (beta-carotene becomes vitamin A in our bodies) located in Freeport, Texas. It's releasing similar toxins to Roche's Belvidere plant, though without much fuss, since its neighbors in the heavily polluted chemical epicenter of Freeport are the likes of Dow, BASF, and Conoco Philips.

Eager to talk to Chinese vitamin manufacturers, I traveled in the fall of 2011 to Las Vegas's SupplySide West, where a dozen of them were scheduled to appear. A bit like the IFT trade show, only smaller and with a greater emphasis on strange nutritional supplements instead of strange food ingredients, SupplySide is a four-day fiesta of the sorts of potions and elixirs that line the shelves of GNC and The Vitamin Shoppe. At the Sands Expo and Convention Center, which is attached to the absurdly baroque Venetian hotel, there were people selling exotic herbal remedies, muscle-boosting amino acids, and various bedazzling powders that may or may not improve your memory, give you more energy, stimulate your libido, eliminate joint soreness, blast belly fat, and improve digestion.

Just as at IFT, big, multinational companies such as Tate & Lyle, Danisco, and BASF pitched enormous, colorful structures. The Chinese vitamin companies, in contrast, had simple, Lilliputian encampments containing only the most basic information about their products. They handed out thin brochures in Mandarin and English, the contents of which illuminated nothing. Some booths featured pictures of the company's plant and a list of the products they offered; others had a small table and chairs where people could sit down and talk. The major manufacturers of vitamin C were there, and when I asked how their products were made, I got the same answer every time. "Fermentation," they chimed.

"What's being fermented?"

"Corn," they declared.

It turns out that "fermentation" and "corn" explain the origins of vitamin C the way "the supermarket" answers the question "where do hot dogs come from?" As I came to realize, the Chinese abandoned chemical synthesis in favor of fermentation, a different but by no means simpler endeavor. Vitamin C production is a convoluted operation consisting of about ten steps, within which there are substeps. It starts not with corn kernels or even cornstarch, but sorbitol, a sugar alcohol found naturally in fruit and made commercially by cleaving apart and rearranging corn molecules with enzymes and a hydrogenation process.

Once you have sorbitol, fermentation starts, a process that tends to muck up surrounding air less than chemical synthesis (although it's been known to cause problems with water pollution). The fermentation is done with bacteria, which enable more molecular rearrangement, turning sorbitol into sorbose. Then another fermentation step, this one usually with a genetically modified bacteria, turns sorbose into something called 2-ketogluconic acid. After that, 2-ketogluconic acid is treated with hydrochloric acid to form crude ascorbic acid. Once this is filtered, purified, and milled into a fine white powder, it's ready to be shipped off as finished ascorbic acid, mixed with other nutrients, and added to your Corn Flakes.

"Fermented corn" is certainly an easier answer.

Vitamin C manufacturing—and the making of B2 and B12, which similarly employs fermentation of corn derivatives and uses GM bacteria—is about as food-based as vitamins get. Other vitamins come from sources not the least bit edible. Vitamin D production, for instance, requires numerous industrial chemicals to turn that Australian (and sometimes New Zealand or Chinese) sheep grease into the nutrient that gets added to milk. Vitamin B1, a nutrient that Splenda consumers might associate with a healthy metabolism, starts with chemicals you get from coal tar, which is just as it sounds—a thick, fragrant brown or black goop produced when a certain type of coal is heated at temperatures approaching 2,000°F. After that come fifteen to seventeen steps of advanced chemistry.

At its B3 plant in Guangzhou, a mainland port near the island of Hong Kong, the Swiss company Lonza starts with a chemical that's a waste product in the production of nylon 6,6, a fabric used for carpets and vehicle air bags. Vitamin A comes from lemongrass oil compounds and acetone, the active ingredient in many nail polish removers, in what the Dutch company DSM refers to as "a highly complex, multi-step process."

Although these vitamins start with unfriendly materials and originate from factories you wouldn't want to live anywhere near, the end result is not only thoroughly edible, but can be quite healthy, particularly if you're deficient in a certain nutrient. You wouldn't want to put coal-tar chemicals in your coffee, but by the time the makers of Splenda and Special K get a hold of them, their hydrocarbon molecules have been safely rearranged into the salubrious substance our body recognizes as vitamin B1.

Or at least that's how it's supposed to work.

The Equivalency Question

The belief long governing food fortification is that with only a few exceptions, a vitamin is a vitamin. It doesn't matter whether they were manufactured by Mother Nature or in a factory in Shenyang. Were you to gaze under a microscope at the two versions of an ascorbic acid molecule, for instance, you'd find they look exactly the same.

But while studies routinely show that various fruits and vegetables are effective in warding off modern ailments like cancer and heart disease, the same can't be said for synthetic vitamins. According to a whole spate of recent studies, most of which were done on pills, not fortified foods, supplemental vitamins fail to offer the health benefits we once thought they had. In some cases, taking large amounts of these compounds appears to have exactly the opposite effect, increasing your chances of getting diseases like cancer. In one study, men who took high doses of synthetic vitamin E and selenium had a slightly higher risk of developing prostate cancer than men who didn't, leading some researchers to suggest that large amounts of vitamins might actually have the unintended consequence of feeding tumor cells. (One study in 2012, however, bucked the trend and showed that standard multivitamin pills lowered mens' risk of all cancers.)

The reasons for vitamins' lack of a clear-cut health benefit still aren't clear, but may have something to do with the fact that we're already consuming enough, or even too many, vitamins. You don't hear this very often—in fact, you're more likely to hear the opposite—but most of us are actually getting our fill of most essential nutrients.

This was one of the illuminating findings of a 2011 study in the *Journal of Nutrition*. A team of researchers led by former Kellogg's executive Victor Fulgoni III looked at Americans' total vitamin and mineral intake from all sources—whole foods, fortified foods, and vitamin supplements. They found that most of us (90 percent or more) are getting the recommended daily allotments of eleven nutrients: zinc, phosphorus, iron, copper, selenium, B1, B2, B3, B6, B12, and folic acid. A majority of us are also consuming enough vitamins A and C, as well as calcium and magnesium—66 percent, 75 percent, 62 percent, and 65 percent of the population respectively. The only things we're particularly lacking in are potassium and vitamins D, E, and K.

A few of us are even getting too many vitamins, something that, at the very least, places an excessive burden on the kidneys. This is of particular concern in kids, who need fewer nutrients than adults and whose kidneys are more susceptible. Some 24 percent of children are going over the upper limit for zinc, 16 percent for B3, and 15 percent for vitamin A and folic acid. An upper limit is defined as the highest level of daily intake that's likely to pose no risk of adverse health.

And here's the amazing part in all this: for all but two (vitamins B3 and K) of these eleven essential nutrients, a majority of them are coming from Fruity Pebbles, Clif Bars, vitaminwater, enriched bread, margarine, and multivitamin pills, instead of eggs, spinach, and whole-wheat toast. In other words, we're now getting more synthetic nutrients than naturally occurring ones. In Europe, the birthplace of synthetic vitamins, the situation is reversed. There, the *Journal of Nutrition* study noted, vitamins and minerals inherent to foods serve as the major source of essential nutrients. The conclusion that Fulgoni, who runs a nutrition consulting firm in Battle Creek, drew from all this data is how critically important added vitamins and minerals are. "Health professionals must be aware of the contribution that enrichment and/or fortification and dietary supplements make to the nutritional status of Americans," he wrote.

There's no question that when you're deficient in some particular nutrient, vitamins, in whatever form, work wonders. In poor areas of the U.S. South during the nineteenth century and early parts of the twentieth century, people routinely developed a disease called pellagra due to lack of vitamin B3. Symptoms include skin lesions, hypersensitivity to sunlight, and mental confusion. In northern climates, it was common to see children with severely bowed legs and other skeletal deformities attributable to rickets from inadequate vitamin D and possibly calcium. Today in the developing world, these vitamin deficiency diseases are still prevalent. In the United States, though, pellagra, rickets, beriberi (lack of vitamin B1), and scurvy (lack of vitamin C) have all but vanished as major health concerns since at least the 1960s, thanks primarily to the fortification of white flour and milk and the widespread availability of supermarket items like orange juice. (Alarmingly, incidence of rickets has reemerged recently, particularly among African-American children who drink more soda than milk. People with darker skin don't absorb as much of the sun's rays, and thus don't produce vitamin D as effectively in their skin.)

So if most of us are meeting the bulk of our vitamin and mineral quotas and yet half of us are suffering from one or more chronic diseases (not including obesity), what's the point of all these factory vitamins?

They don't seem to be helping us live longer, at least compared to our global peers, who don't eat as many fortified foods. Americans rank dismally for life expectancy at number 37, behind nearly all Western European nations. We barely eke ahead of Slovenia. Americans have also lost the first-place trophy for tallest population on the planet, having consistently slipped in the global height rankings for the past half-century (although we're still much taller than our nineteenth-century forbears).

Does this mean that the synthetic vitamins and minerals we're consuming by the fistful are not, after all, equivalent to the naturally occurring ones in comparatively short supply in our diet? "It's a good question," said Marian Neuhouser, a nutrition researcher at the Fred Hutchinson Cancer Research Center in Seattle, when I posed it to her over the phone one day. "There really haven't been studies done comparing blood values and metabolic values of vitamins from synthetic or nonfood sources versus whole foods."

That we wouldn't know the full story of something as significant as vitamins and minerals struck me as surprising. It's as if scientists, having put unwavering faith in the marvels of modern chemistry, decided that the matter of synthetic vitamins was settled long ago. But the more we learn about the nutritional complexity of whole foods, the more the issue appears far from settled.

Scientists are only beginning to understand the many ways in which nourishment is a complex symphony of not just vitamins and minerals but also phytochemicals, enzymes, fiber, and gut bacteria. Beta-carotene, for instance, never appears just by itself in fruits and vegetables. It's part of a family of other carotenoids. So, too, with vitamin E. The form added to food is alpha-tocopherol— only one of eight compounds comprising the vitamin E you get in sunflower seeds and peanut butter. So do we really reap the same benefits when vitamins are manufactured as when we eat them with a concert of other nutrients found in whole foods?

The newest research suggests we do not. In 2000, a researcher at Cornell did the sort of enlightening study that wasn't possible when vitamins were first synthesized, nor even as recently as two decades ago. He looked at the antioxidant activity of apples and found that ascorbic acid, or vitamin C, is responsible for only a small portion of this important effect. Most of the fruit's antioxidant power comes from other phytochemicals, some of which work in tandem with ascorbic acid. It's a finding that casts doubt on the many antioxidant boasts on food packages. If vitamins need these other things to work effectively as antioxidants, then the antioxidant benefits of, for instance, Kellogg's Fiber Plus Antioxidants cereal, based solely on the addition of vitamins C and E, or of Diet Cherry 7Up Antioxidant (with vitamin E), are likely to be miniscule.

"A lot of people have started to wonder if the whole is more than the sum of its parts," said Neuhouser.

The fact remains that we simply don't know all the parts. Not that we aren't trying.

Phytomins

Several years ago Shengmin Sang mapped out, for the first time, the specific components in wheat bran that are responsible for its ability to help prevent colon cancer. He and his research team did this in their shiny, immaculate lab in North Carolina, using human colon cancer cells and advanced liquid chromatography machines, devices that cost half a million dollars and look like they might have been transported from Buck Rogers's spaceship. The whole project took about a year and challenged the notion that fiber is the primary reason wheat bran is effective for colon cancer prevention. Sang's project showed it was the work of fourteen thoroughly healthy substances, some of them previously unknown, that he identified with names like 7, (10′Z,13′Z,16′Z)-5-(nonadeca-10′,13′,16′-trienyl) resorcinol.

Scientists have long suspected that food's nutritional story has a lot more to it than just vitamins, minerals, protein, and carbohydrates. It's now understood that phytochemicals—substances that all plants produce in part to ward off pathogens like bacteria and fungi—are also an important, if still mysterious, piece of the puzzle. They help account for the fact that fruits and vegetables are universally regarded as healthy. Has anyone ever argued against the healthiness of broccoli?

These curious chemicals (which include polyphenols, carotenoids and flavonoids) may hold the keys to slowing down the aging process, keeping blood pressure levels normal, and preventing the conditions that lead to heart disease and cancer. And while they're currently not considered essential the way vitamins and minerals are—meaning they're not believed to be absolutely necessary for proper growth and functioning—they might be one day. Joe Vinson, a chemist at the University of Scranton, calls phyto-

chemicals "the bioactive components in foods, more so than vitamins because they have more mechanisms associated with them."
Louisiana State food scientist John Finely says the bottom line for
now is that "we're far better off with them than without."

Yet there's much we still don't know about these substances.
The state of the science is roughly on par with where the knowledge of vitamins was in 1920. No one is quite sure, for instance,
exactly how many of them there are. Some 10,000 have been identified, yet many of these remain an enigma. Because it's not enough
to simply dissect the structure and properties of the phytochemical itself. Scientists also want to figure out how it's metabolized by
the body and what the health effects of those various metabolites
might be.

People like Sang are working on just that. A slight, soft-spoken
chemist whose angular features and intense gaze can make him
look like he's frowning when he's not, Sang grew up in Shandong,
China. After getting a chemistry PhD, he started looking for jobs
the way resourceful modern graduates do—on the Internet. He ultimately found one at Rutgers University as a postdoctoral research
associate in the food science department. At $27,000 a year, it paid
a meager salary for someone with a PhD, but Sang eagerly took it.
"It's hard to get people here in the U.S. to do postdoc research," he
explained when we met up at a chemistry conference, where he was
accepting an award. "Most Americans would prefer to have a job
in the industry, which pays a lot more, so you have to get people
from overseas."

Sang's prolific work at Rutgers—more than a hundred published, peer-reviewed studies in ten years—helped him land his
current job at the North Carolina Research Campus (NCRC), an
ambitious, $1.5 billion nutrition and health complex in Kannap-
olis, twenty-five miles outside Charlotte. Built in 2008, the NCRC

is a collaboration among the area's seven universities and the brainchild of an exceptionally healthy, eighty-nine-year-old billionaire named David Murdock. The former CEO of Dole Foods, Murdock founded each of the half dozen institutes at the complex on the belief that fruits and vegetables hold quite a few of the keys to human health and longevity. Murdock himself is determined to see the north side of a hundred with unclogged arteries and limber joints.

Some programs at the complex work to improve plant nutrition through breeding; General Mills is funding research with the goal of developing oats with higher and more consistent levels of the cholesterol-reducing fiber known as beta-glucan. Other efforts are aimed at optimizing plant harvesting and storage and studying the link between genes and diet. And there are initiatives, like Sang's, to unlock the secrets of phytochemicals such as polyphenols.

In addition to wheat bran, Sang has also studied apples, ginger, rosemary, tea, and blueberries. He's shown that pterostilbene (pronounced *tero-STILL-bean*), one of the factors giving blueberries their antioxidant superfood power, breaks down into nine different compounds in mice, some of which may or may not have their own health benefits. He found that various bioactive components in apples and substances called catechins in tea can trap compounds thought to be involved in the formation of type II diabetes.

It's the sort of work that food companies are watching with great interest. The beverage industry is particularly intrigued with the idea of phytochemicals as the new generation of vitamins and minerals. In 2011, Coke funded a study in Britain testing the effects of a juice drink containing a slew of exotic compounds from green tea, grape seeds, apples, and grapes. Researchers at the University of Glasgow fed this superjuice to thirty-nine middle-aged, overweight, but otherwise healthy people, and two weeks later, tested their urine. They found that the superjuice group had significant changes in various biomarkers associated with cardiovascular disease, as compared to a control group of people drinking boring old juice. "The profile of compounds present changed toward that of a healthier individual," explained Bill Mullen at the University of Glasgow, in an e-mail. Coke, for its part, said the project will help them evaluate the effectiveness of polyphenols—both those added to its juice products, such as is the case in Europe, and those that may exist naturally in the company's juices.

When I asked Sang what he thought about phytochemicals being used in juice drinks or other processed foods, he was of two minds about the possibility. Like vitamins, many phytochemicals are destroyed or removed in manufacturing and therefore aren't particularly abundant in processed foods. Adding them back in, he said, wouldn't work "from a biological point of view," meaning they might not function effectively when isolated from their natural fruit and vegetable habitat. Also, if juice drinks have high amounts of added sugar and if other processed foods contain potential carcinogens like nitrites, he said, added antioxidants may not generate much in the way of health benefits. "Americans think you can eat poorly and then take supplements," he said. "The idea is to try your best to eat healthy in the first place."

Sang himself appeared to be quite healthy. Inspired in part by his research, he eats two apples a day and drinks only water and green or ginger tea. "I avoid processed meat and eat only a little chicken and fish," he said, adding that he has yet to find a proper Chinese restaurant in the Kannapolis area. His chief personal motivation for deconstructing plant foods (which, when you think about it, are just fine the way they are) seemed to be the identification of ways that these foods—not the isolated, manufactured compounds—can be used to help fight disease.

Although Coke's juice study showed benefits from a combination of half a dozen different phytochemicals, other studies have suggested that these materials don't work very well in their extracted form. They need a whole slew of other things to help them live up to their potential. Emily Ho, a nutrition scientist at Oregon State University, looked at this in two studies she did involving broccoli. After feeding a group of people fresh broccoli sprouts and then a month later giving them pills containing glucosinolate—anticancer broccoli compounds—she found that the subjects absorbed and metabolized seven times more of the phytochemicals from the broccoli sprouts than from the supplements.

What was missing in the extracts, Ho said when I spoke to her over the phone, is a naturally occurring enzyme in broccoli that helps turn the glucosinolate into sulforaphane, a substance that's been found to help detoxify carcinogens and activate genes that suppress tumors. And it may be even more complicated than that. "At the cellular level, there's the question of whether combinations of compounds have synergistic effects inside the cell, working on different pathways or helping each other to cause a biological effect," she explained. John Erdman at the University of Illinois arrived at similar conclusions about powder from whole tomatoes versus lycopene. "The hypothesis that whole foods are always going to be better is a good one," he said.

There's an ocean of difference between actual fruits and vegetables and the phytochemicals that may appear one day in products at the supermarket (and that are already sold as supplements at GNC). Like vitamins, they will not be derived from their edible hosts but produced synthetically, the vastly more efficient and economical route. Most companies selling lycopene as a nutritional supplement, for instance, are making it from chemical synthesis, not prying it from tomatoes. The blueberry compound pterostilbene, advertised on TV as "BluScience" and thought to lower cholesterol and slow the mental decline that comes with age, is constructed from chemicals such as dimethoxybenzylphosphonate and tetrahydrofuran. According to Chromadex, the Irvine, California-based company producing it, the job of extracting just 50 milligrams of pterostilbene from blueberries would require anywhere between 200 and 700 pounds of fruit, which, given the steep price of blueberries, would make the whole operation unfeasible.

Yet at the same time Sang acknowledges the limitations of such added phytochemicals, he realizes there's a gathering interest in the food and beverage industry in products with specific health benefits, or so-called "functional foods." There's no money to be made in publicizing the fact that apples are healthy, but the commercial appeal of a blueberry cereal bar with blueberry antioxidants or a multigrain bread with anticancer wheat-bran compounds is obvious and compelling. Not long ago, General Mills said it was looking for interesting new healthy ingredients that it could incorporate into its products, particularly breakfast cereal. These additions would be substances with a proven ability to lower risk factors for heart disease, to enhance mental functioning or to boost satiety.

Sang's official biography states that the goal of his research is to do just this sort of thing—identify "novel bioactive natural products that can be used in functional foods." So it's not really surprising that, despite his misgivings, he seemed willing to reconsider his thoughts on how phytochemicals should be used. Toward the end of our conversation, he said that if Coke or PepsiCo were to put a flavonoid from green tea, say, into one of their drinks, even sodas, that could be okay. It's not ideal, but with enough research there might be a way for it to work. A placebo effect, at the very least. "It might be better than not having it," he mused.

It was a pragmatic thought, but not one that offered much reassurance about the ability of food scientists to engineer products anywhere near as healthy as the real thing. Humpty Dumpty, it would seem, doesn't come together all that easily.

6

Better Living through Chemistry

Personally I stay away from natural foods. At my age
I need all the preservatives I can get.

—George Burns

Several weeks after that April evening when my mom accidentally ate the nine-month-old guacamole, I decided to try and
track down the origins of those mysterious ingredients—the "text-
instant" and "amigum." She hadn't been harmed by them, but I
wanted to get to the bottom of why a seemingly simple food like
guacamole needed such things. I contacted a food scientist who's
held a wide range of jobs in the industry and has worked with all
kinds of ingredients. He, too, had never heard of "text-instant"
or "amigum." It piqued his interest, and before I had a chance to
do it, he contacted Kroger, the grocery chain where my husband
bought what had become a very interesting tub of guacamole.

Within a day, he'd gotten an e-mail response from Molly

McBride, a Kroger corporate dietitian, who had inquired of the company's food scientists. She explained that "text-instant" was a mislabeled ingredient. Instead, it was modified starch. Labels, she said, would be corrected to read "food starch modified." The food scientist took this to mean that it was probably a corn or tapioca product from Ingredion (formerly National Starch) that, as the company puts it, gives products "pulpy textural characteristics." The product's called Instant Textaid.

McBride made no reference to "amigum" in her e-mail, only to indicate that it, too, would no longer appear on the ingredient label. The food scientist found this lack of explanation curious. The only amigum he knew of was something used in cosmetics as a gelling agent, and for which the safety data sheets read: "If ingested, obtain medical attention." He had a disquieting theory: "I think their vendor maybe took a basic avocado facial mask formula and turned it into a dip without realizing that one of the ingredients was not approved for food use."

So there you have it: My mom, forever wary of "gooped-up" products, may have unintentionally consumed facial mask.

As for the absence of mold or other red flags indicating that this was a nine-month-old specimen—that's thanks to the ascorbic acid and citric acid, two ingredients originating from corn. Or something like corn. Both additives lowered the guacamole's pH, reducing it to the point where no fungi, yeast, or bacteria could grow. Adding lime juice will accomplish something similar, though to a lesser extent. Even with lime, homemade guacamole will turn brown after five days, and within a week or so, without the binding effects of modified starch and gels, water can leak out, forming a moat around the edges. And when you see snowflakes of white mold dotting the surface, there's no question it's time to make a fresh batch.

Citric acid, ascorbic acid, and Instant Textaid (but not amigum) are three among some five thousand additives going into our food. To put this number in perspective, it's helpful to realize that this is more substances than we've ever consumed before. While the practice of doctoring our food is nothing new—salt, the original food additive, has been used for thousands of years to keep food, especially meat, from going bad—the sheer number of things now being used to do it certainly is. The story of how we got here is at once a tale of lax regulation and an account of the often surprising extent of manipulation our food undergoes.

Besides improving texture and making food last longer than it has any right to, food additives help make food taste good, either by adding desirable flavors or masking those that might otherwise send us reeling in disgust. They make everything stick together and give what food scientists call good "mouthfeel." They also help food look attractive, like the Yellow #5 that brightens up the Kellogg's Smart Start cereal pellets emerging drearily gray from extruder machines. Or the propylene glycol—a nontoxic antifreeze chemical and ingredient in BP's oil dispersant in the Gulf—residing in the "California Natural" honey cashews I bought not long ago at an airport. It was there to prevent the honey/sugar glaze from darkening.

And there are other functions for additives, including ones you'd never imagine and for foods you might not think would need them. Foods such as bread.

Why Bread Needs Strange Ingredients

All that's really needed to make this five-thousand-year-old staple is four ingredients—flour, water, salt, and yeast. Many recipes will call for some type of sugar and a fat, and you can get fancy with spices, dried fruit, nuts, and seeds. But at base, the four-ingredient approach is more or less the way bread is still made in many places around the world—in the small French *boulangeries* that gave the world the baguette, the state-subsidized bakeries in North Africa that families visit daily, the German shops that produce bread in endless varieties, and artisanal bakeries here in the United States.

This simple, ancient approach is unfortunately not well suited to modern modes of mass production. It's time-consuming, mostly because of the need for dough to ferment and rise, and it produces a relatively dense—hence expensive—loaf. Most large American bakeries long ago abandoned traditional bread making in favor of much lower-cost forms of production, ones that require a lot more than four ingredients—usually quite a lot more. Rather than a basic staff of life, commercial bread, both in the supermarket and at chain restaurants, is an edible testament to the triumph of chemical technology.

Subway's bread is a great example of this. Inside Subway stores, tubes of bread are baked several times a day, emerging, as all recently baked bread does, as a soft, doughy temptation. Its distinctive and yeasty aroma permeates stores and evokes memories of homemade bread (even if largely imagined memories). Notwithstanding what is sometimes an unsettlingly complex smell, such in-store baking, said Mark Christiano, Subway's global baking specialist, is "part of the romance of what we do." If you define age from the time the bread exits the oven, then at Subway "you're going to get a product that's only a few hours old at the most," he said. "We're always baking fresh bread. The customer sees it, and we're very proud of it."

It was back in 1983 that Subway decided to stop buying the prepackaged bread everyone else was using and instead buy ovens, putting them right at the front of the store. It was a brilliant move that predates the popularity of Panera Bread, which proudly, and somewhat misleadingly, touts its in-store bakers on the sides of its delivery trucks. Neither Panera nor Subway go through the trouble of making the actual dough in their stores. Panera ships it daily from regional baking centers—all the celebrated in-store "bakers" have to do is pop it in the oven—and Subway stores get it frozen from one of ten large, contracted factories around the country.

Over the past decade, Subway's bread has formed the cornerstone of the chain's hugely successful "Eat Fresh" campaign. Some combination of this effort, an obsessive promotion of low-calorie/low-fat options, and the compelling story of Subway's once-obese spokesperson Jared Fogle has served to catapult the chain to the top of the rankings for "healthiest fast food." In 2011, the company was the only one to get an A in *Men's Health*'s silly *Eat This Not That* restaurant report card. They were also identified as the most popular mega chain by Zagat's. "[Customers] like the freshness of the ingredients, particularly," Tim Zagat told the *Today Show*'s Matt Lauer. But if you actually look at them, these ingredients, or perhaps the ingredients within the ingredients, which number between twenty and fifty for the bread alone, tell a different kind of story, one that doesn't have much to do with freshness or health.

When subjected to industrial production, bread isn't so much mixed as beaten up. Inside massive rectangular high-speed machines, it's churned and pummeled, forcing a lot of air and water into the dough. This lowers costs by boosting the amount of bread you can make from every pound of flour. The turbo-tossing also moves dough through assembly lines more quickly, reducing the amount of time needed for it to rise. The result is something containing less bread and more air, and a soft, fluffy texture many Americans have come to expect. When I was in elementary school, I remember kids taking slices of Wonder bread, balling them up, and flinging them across the cafeteria when the nuns weren't looking. Although Wonder bread no longer dominates the nation's lunch boxes as it once did, you can still turn slices from brands like Sara Lee and Oroweat into spherical weapons. A six-inch tube of Subway's bread can easily be balled into the size of a lemon.

In order to get dough to survive all this puffing up and thrashing that happens inside machines, "dough conditioners" are needed. Subway's bread contains five of them: sodium stearoyl lactylate, monoglycerides, diglycerides, ascorbic acid, and diacetyl tartaric ester of monoglyceride, known as DATEM. Without these ingredients, Subway's dough would break down, losing its elasticity and sticking in gooey clumps to the machines. These additives, which are used in most fast-food and supermarket varieties as well, infuse bread with a quality most traditional bakers have probably never thought to consider—what baking scientists refer to as "dough machinability."

Calling dough conditioners "chemicals" isn't entirely accurate, because they start with a fat, such as soybean oil, beef tallow, or palm oil. After that, you need a heavy-duty floodlight to guide your way down the industrial rabbit hole. Here's a quick glimpse: To create sodium stearoyl lactylate, the fat is first sheared into its component fatty acids in a process known as thermal cracking, something commonly employed in petroleum refining. The resulting stearic acid or palmitic acid is then combined with heat, lactic acid, and sodium. That gives you the *sodium stearoyl* part. To get lactylate, cornstarch is treated with enzymes to yield dextrose, which is then fermented to produce lactic acid. Monoglycerides and diglycerides come from applying an alcohol and an acid to glycerin, a derivative of soybean or palm oil. Further processing of these compounds with tartaric acid, a naturally occurring chemical found in wine though derived synthetically, yields DATEM. Sodium stearoyl lactylate and mono- and diglycerides are not exclusive to bread. They're found in lots of other foods, in categories ranging from frozen meals and desserts to breaded meat products, pancake mixes, kids' lunches, and Starbucks's apple fritters and pound cakes (despite the chain's emphasis on natural ingredients).

One of Subway's other bread ingredients—azodicarbonamide—is used in the chain's white and sourdough breads to help evenly distribute the air pulled in from mixing. This gives loaves the appearance of neatly woven pieces of fabric, with each tiny air pocket more or less the same size—a work of perfection that bakers call a "fine crumb structure."

A yellowish orange powder, azodicarbonamide is produced from hydrazine, which comes from a reaction of the chemicals sodium hypochlorite and ammonia. Its biggest uses are not in other areas of the supermarket but in the production of rubber and plastics. Azodicarbonamide has probably been used to make the soles of your shoes and the floor mats you walk on at the gym. It's not something anyone would ever think to eat. On a summer morning in 2001, a truck carrying it overturned on Chicago's busy Dan Ryan Expressway, prompting city fire officials to issue the highest hazardous-materials alert and evacuate everyone living up to a half a mile downwind. People on the scene, many of whom abandoned their cars amid the massive traffic pileup, complained of burning eyes and skin irritation.

Making mass-produced bread without flammable chemicals and highly engineered fat derivatives is possible; it's just considerably harder and likely more expensive. Subway's lengthy list of bread ingredients ensures that production is pretty close to foolproof and that the small numbers of people employed in their automated factories don't have to contend with too many x factors during manufacturing. It also guarantees that once baked, the loaves look, taste, and feel the same every time, in every location.

A certified organic baker I talked to recalled, somewhat mournfully, the toolbox of synthetic additives he relied on during his thirty years working for big, conventional baking companies. "You have your SSLs, your mono and di's, your ADA [azodicarbonamide], your cal pro's [the preservative calcium propionate], they all make it easier," he said. "With organic, you have to be spot-on." None of these ingredients are allowed in organics, and he showed me how as a result his bread—Rudi's Organic, based in Boulder—has occasional air pockets and other minor imperfections. And it costs more, selling for more than $4 a loaf, versus a national average of between $2 and $3, though a good deal of that has to do with the higher cost of organic wheat. Great Harvest bread, which is even closer to traditional bread making, featuring hand-kneading and on-site fresh grinding of wheat, is even more expensive, averaging more than $5 a loaf. But when you taste this incredible bread and consider what's not in it, the cost might be worth it.

Generally Recognized as Safe

The FDA is the government agency responsible for overseeing the safety of all our food except fresh meat (that falls to the USDA). It's a monumental responsibility, and the agency often seeks to convey the impression that the large number of materials going into our food supply are under close watch. "All food additives are carefully regulated by federal authorities and various international organizations to ensure that foods are safe to eat and are accurately labeled," the FDA states on its Web site. Yet the truth is nowhere near as reassuring. The food industry's blistering pace of innovation and the force of its lobbying efforts have always overwhelmed those charged with reigning it in.

Despite Harvey Wiley's success in helping to usher in food legislation, the numbers of new chemicals and additives going into food exploded in the years after he left government. An influential book published in 1933 entitled *100,000,000 Guinea Pigs* referred to America's food supply as a "curious, one-sided experiment with bleaches, preservatives, adulterants, fillers and poisons—and with men, women and children as test animals in place of rats and guinea pigs." The book quoted a top-ranking FDA official who was, even then, frustrated by the impossibilities of his job. "Adulteration of food products," he lamented, "is almost invariably a step or two in advance of the regulatory food chemist." In the mid-fifties, when Representative James Delaney from New York sat down to tally up the number of substances going into food, the sum had reached roughly seven hundred. The legislation Delaney helped pass in 1958, the Food Additive Amendment, chopped this number down to about one hundred and eighty eight, though from there it regrew, salamander-like, at an even faster rate than before, thanks to the boom in new processed foods. By 1980, there were some two thousand additives allowed in food, and the list has only expanded from there.

The truly astonishing thing about today's five-thousand-additive figure is that it comes not from the FDA or some other government body, but from Pew Charitable Trusts, a nonprofit group that did a first-rate report on food additives in 2011. To arrive at their number, an estimate, Pew accounted for ingredients from multiple layers of a convoluted maze of food-additive regulation. Amazingly, the FDA itself has never attempted this exercise, at least not publicly. If it knows the sum total of ingredients going into our food, it's not letting on.

There's no government Web site, for instance, you can go to that will list all those five-thousand-plus substances. There's a database called Everything Added to Food in the United States (EAFUS) and you might assume this would do the trick, but not only does this absurdly titled compilation not include everything that's added to food, it contains no real data on the substances it does list. According to Pew, EAFUS incorporates less than half of all food materials and less than 10 percent of the new additives that have gone into our food in the last ten years. Another database, the Food Additive Status List, isn't any more helpful, especially if you're looking to understand what a food ingredient actually is and what studies exist on its safety. Here, in its entirety, is the illuminating entry for azodicarbonamide, which appears not just in Subway's bread but that of McDonald's, Burger King, Arby's, Wendy's, Dunkin Donuts (croissants, English muffins, Danishes), and Sara Lee:

BL, REG/FS, 45 ppm in flour—Part 137, Cereal Flours & 172.806

And nowhere on any site will you find mention of what Pew has estimated are roughly a thousand ghost additives out there on the market. These are ingredients companies have simply declared safe on their own without so much as an e-mail to the FDA. Pew says that the actual number of these additives is unlikely to be less than five hundred but could be several times larger than one thousand.

Grasping how the FDA could allow companies simply to wave a magic wand and give their own ingredients a green light requires going back to that 1958 food-additive law. In passing it, Congress expected that all new substances would go through a rigorous FDA review process before being launched into the supermarket. Congress wasn't sure how many new food additives there would be, but it naively figured the total might reach a thousand or so. As a sort of side rule, the law established another program known as the GRAS list (Generally Recognized as Safe) for substances that everyone and their dog knew to be safe—things like spices, salt, vinegar, and yeast. It was an escape hatch for the relatively small number of commonsense things that didn't need an arduous approval process. And thanks to the compromise-ridden, sausage-making nature of government legislation, it was made voluntary. Companies didn't have to let the FDA know about their GRAS ingredients if they chose not to, though Congress expected most of them would anyway.

Ingredient companies quickly realized that getting something declared GRAS by the FDA was infinitely easier than vying for approval through what was formally and confusingly called "the food additive process." Although it was never intended to, the torrent of new ingredients gradually shifted toward GRAS.

Then in 1997, a change in the rules made getting something on the GRAS list even easier, turning the stringent food-additive process into a vestigial organ of the regulatory system. The FDA said that instead of GRAS *petitions*—a filing process where agency scientists had to look at safety data and make a decision—there were now GRAS *notifications,* a system by which a company would assess the safety of its own ingredients, often by assembling a panel of experts. The company would then *notify* the FDA of its decision, unless, of course, it decided not to. The system, remember, is still voluntary. The GRAS process became so attractive to ingredient companies that everyone pretty much stopped using the petition process Congress had set up. Since 2000, there have been formal FDA "food additive petitions" for only four new substances.

In deciding their own fate, companies are supposed to adhere to guidelines laid out in what's known as the FDA's Redbook, but there are no legal ramifications for not following these nonbinding guidelines. What's more, the standards for what sorts of qualifications an "expert" must have were never spelled out beyond the expectation that it should be someone "qualified by scientific training and experience to evaluate [an ingredient's] safety."

I called George Burdock, a toxicologist who's been doing GRAS determinations for twenty-five years, to ask him about this. Over the phone, I could hear the frustration gathering in his voice as he spoke. His firm, he said, seeks to do rigorous evaluations, and sometimes, though by no means always, it recommends animal testing for new ingredients. But not everyone operates this way. He's had more than one client take their business elsewhere because they didn't like what the Burdock Group told them. "There are people out there who will get on Google and do a search for existing research, then tell you the good news," he fumed. "Some of these people have a 'Dr.' in front of their name, but they might be a doctor of divinity. It's like when you used to see law degrees advertised on matchbooks."

Tom Neltner, the chemical engineer who's leading Pew's food-additive project, told a similar story. He's counted eleven instances out of 410 of an expert "panel" of one and knows of a single person who has served on 185 different panels. "That person may have done a wonderful job and he may have been picked because he's the most responsible, but it raises a lot of questions," said Neltner. The FDA has the ability to respond to and effectively derail such paltry evaluations, but when it does, companies are first given the opportunity to withdraw their notification. Neltner identified thirty-one such withdrawals since 2001. What's allowed to happen next is even more remarkable. Companies can either drop the ingredient (unlikely), do additional investigations in order to resubmit to the FDA, or simply go the abracadabra, ghost-additive route—declare the item safe and start selling it. Pew has found numerous instances of this.

Pew's five-thousand-ingredient figure is just an account of the substances added directly and intentionally to food. It doesn't include what are known as food-contact substances—things manufacturers use in their packaging and apply to machinery to keep it running, such as lubricating oils and cleaning chemicals. There are 3,750 of these substances, though companies never intend for them to migrate into our food. Often they do, but you only hear about it when there's a problem. Like the time customers called Kellogg's toll-free phone number complaining of nausea, diarrhea, and a noxious, waxy odor emanating from cereal boxes. The cause turned out to be elevated levels of an unstudied chemical called methylnaphthalene that's a breakdown product of various chemicals used in the wax-lined cereal bags. In July 2010, Kellogg issued a recall for 28 million boxes of Froot Loops, Apple Jacks, Corn Pops, and Honey Smacks but never explained what went wrong, other than to say that the substance in question (it never identified methylnaphthalene; the Environmental Working Group sussed it out) is used commonly in packaging materials and in resins to coat foods such as cheese, raw fruits, and cucumbers, an explanation that doesn't quite inspire confidence. It's a small miracle this sort of contamination doesn't happen more often, considering the many chemicals that can function as food-contact substances in a single product. In addition to methylnaphthalene, the production of Kellogg's Froot Loops could include as many as twenty other food-contact substances.

Fox in the Henhouse

The chief problem with food-additive self-regulation is that it provides no incentive for either the industry or the government to do—or even to review—the difficult and often expensive studies that might result in bad news. Nor does the current system allow for any level of transparency, which is something the 1958 Congress clearly intended. The formal additive petition process requires public hearings, the opportunity for people to file comments, and the means for decisions to be challenged. None of this happens with GRAS.

So despite assurances from the agency and the industry, it's impossible to really know whether all those five thousand substances are safe for us to be eating regularly. The startling reality is that less than half of all food additives (including the indirect food-contact ones) have been the subject of any published toxicology studies, which means feeding them to either animals or humans (mostly animals). In other words, 50 percent may have been deemed safe purely on the basis of the fact that other similar chemicals are already on the market, substances that themselves could have been deemed GRAS on the backs of other compounds. Neltner calls it a "daisy chaining of chemicals."

The last time the FDA initiated a complete review of the scientific status of GRAS ingredients was in 1972. Nixon ordered it in 1970 following an uproar over and delisting of the artificial sweetener cyclamate. Of the 415 substances studied, twenty-five were recommended for restrictions or removal from the list, and nineteen others were flagged for more study and reevaluation. Another 68 were deemed safe at current levels, but in need of more research on whether prolonged use might be a hazard.

Follow-up has been scarce. Were the FDA today to conduct a thorough review of the five thousand substances, more than a few ingredients might not make the cut. The preservative BHA, for one. Among those ingredients singled out for more study in the seventies, BHA—short for butylated hydroxyanisole—is on California's Proposition 65 list of cancer- or birth-defect-causing chemicals. The Department of Health and Human Services, of which the FDA is a part, has placed it on its list—a relatively short one—of substances "reasonably anticipated to be human carcinogens." Because of this, manufacturers have scaled back on its use, turning instead to BHT, which was also on the flagged list but is presumed to be safer. But BHA is still out there. See Tang, tropical punch and lemonade Kool-Aid, DiGiorno pepperoni pizza, and McDonald's sausages and breakfast steak.

Potassium bromate, too, is still legal. The best dough conditioner bakers have ever known, bromate has been used in bread since at least the 1920s. Over the years, various studies have raised concerns, particularly those showing that it causes tumors in rats. The historical argument was that these dangers weren't relevant because the chemical burned off during baking. When tests done in Britain in the late eighties showed that not to be the case, the FDA took action, if you can call it that. In 1991, the agency *asked* bakers to stop using potassium bromate within two years. Lacking a mandate, companies continued putting it in their bread throughout the nineties and early into the last decade. It's only recently become hard to find a loaf or fast-food bun that contains it.

One of the ingredients that replaced potassium bromate—that substance that shut down the Dan Ryan Expressway, azodicarbonamide—may not be that much better. It was recently found to break down into a known carcinogen called semicarbazide when heated.

Health Canada, our northern neighbor's FDA equivalent, was the first to document levels of this carcinogen in bread samples. A year later, the FDA followed suit with its own tests. But instead of asking bakers to stop using azodicarbonamide, as it did with potassium bromate, the FDA suggested that companies please just use a bit less of it. For its part, Health Canada hasn't made any changes to its regulations and believes that both azodicarbonamide and the resulting semicarbazide are safe at their permitted usage levels.

That's also the rationale for TBHQ, an effective antioxidant preservative for fats that restaurants like McDonald's and Arby's use in their frying oils. In animal studies, the chemical has been linked to convulsions, liver enlargement, and precursors to stomach tumors. Yet the FDA allows use of restricted quantities of the preservative and holds that those small amounts are safe.

Even if yellow-orange flammable materials and neurotoxic chemicals were fine to consume in relatively small doses over a long lifetime (a big if), this wouldn't address the question of what happens when these substances are eaten together with others in endless numbers of varying combinations, which is exactly how our bodies receive them. Someone consuming a Nutri-Grain bar in the morning, a Subway Chipotle Chicken and Cheese sandwich for lunch, and a DiGiorno pepperoni pizza for dinner, for instance, will have ingested a total of sixty-eight different nonfood additives (not including vitamins and minerals) that until recently no human being ate. Such amalgamations can't be easily tested; animal studies are rarely done for multiple additives. "Testing chemicals together quickly becomes a 'what if' scenario with an infinite number of possible combinations," noted George Burdock. The best you can do is to run ingredients through software that uses mathematical models to assess the probability of chemicals interacting with each other.

The possibility of problems from chemical interactions isn't just theoretical. In 2006, the beverage industry acknowledged that when ascorbic acid and either sodium benzoate or potassium benzoate were combined in certain conditions, small amounts of benzene, a substance known to cause leukemia and other cancers, were formed. This occurred in products like Sunkist Grape and Orange sodas, Kool-Aid Jammers orange drink, Crystal Light Sunrise Classic Orange drink, and Giant Light Cranberry Juice Cocktail. The solution in many cases was the addition of another chemical that could blunt the formation of benzene—the preservative calcium disodium EDTA. Widely used in sauces and condiments, calcium disodium EDTA (the EDTA part stands for ethylenediaminetetraacetic acid) has been shown to cause kidney damage in laboratory animals, and in sensitive individuals it can result in upset stomach and muscle cramps.

Another overlooked aspect of runaway food-additive use is the possibility of overconsumption. Take phosphate, for instance, an essential mineral that occurs naturally and quite commonly in foods such as meats, grains, nuts, and legumes. We've been consuming it since the beginning of humanity and it's safe, as long as you don't consume too much of it. When found in natural sources, phosphate occurs in a bound form, intertwined with either plant or animal tissue in such a way that our bodies can only absorb 40 percent to 60 percent of it. The phosphate used in food additives like sodium phosphate, on the other hand, is unbound and quite available, according to scientists who've studied the issue. Taking in too much phosphate can place a heavy burden on our vascular system, actually turning muscle cells in our arteries into hardened bone cells, thereby increasing heart disease risk and accelerating the aging process. It can also interfere with the assimilation of calcium. "The phosphate load of today's food, especially in the U.S., is a lot bigger than it used to be with food without additives," wrote Eberhard Ritz, an internal medicine specialist at Ruperto Carola University in Heidelberg, Germany, in an e-mail. "And it's worrisome because the customer cannot see how much phosphate is being added to the food he is buying."

If you start to look, you'll see phosphate-based ingredients everywhere. Kraft's Mac & Cheese, the much-loved kids' meal in a blue box, has a trifecta of this mineral—sodium tripolyphosphate, sodium phosphate, and calcium phosphate. KFC's grilled chicken delivers sodium phosphate and potassium phosphate. If you follow that with one of the chain's strawberry shortcake parfaits for dessert, you'll also ingest monocalcium phosphate, disodium phosphate and sodium acid pyrophosphate. Add a diet or regular Pepsi to the mix (phosphoric acid) and you've got a massive influx of something we only need a little of.

Dihydrogen Oxide and Other Scary Chemicals

It's important to point out that just because food ingredients have unpronounceable names, unappealing origins, and a slippery approval process that's not even an approval, this doesn't immediately qualify them for a skull and crossbones sticker. The industry is fond of bringing up this fact. Fergus Clydesdale, an elder statesman of food science and one of the more outspoken defenders of processed food, gamely explained his view of food chemicals to me. "When people tell me, '*This food contains chemicals,*' I say, I certainly hope so. We'd be in real trouble if it didn't." Clydesdale, a small, wiry, gray-haired man who's been at the University of Massachusetts at Amherst—the country's oldest food science program—for forty-six years, underscored that all foods found in nature are made up of thousands of compounds that can be identified by their chemical names and mapped according to their molecular bonds of hydrogen, carbon, oxygen, nitrogen, and so on. Water, for instance, can be described as "dihydrogen oxide." Were oranges and carrots to be spelled out this way, Clydesdale pointed out, a mass exodus from the produce aisle might ensue. "It would put an end to this fear of what's in food," he said. "They're all very scary sounding chemicals, and some of them really are scary, though they're in there in low amounts."

And then Clydesdale offered an intriguing description of the act of eating. "The job of food," he said, "is to replace chemicals in the body with chemicals in food." When viewed this way, things like phosphoric acid, BHT, and calcium disodium EDTA don't look so bad, since some part of all these substances already exists in our bodies. But Clydesdale's reductionist argument, compelling though it may be, overlooks the essential fact that man-made chemicals are not always the same thing as what's found naturally in foods. Food's chemicals are often packaged differently, come in doses our bodies have adapted to handle, and are found in combinations with other elements that make them work beneficially. Phosphate is just one example.

Without having an advanced degree in food science, it can sometimes be hard to make determinations about what constitutes a healthy food. I remember some years ago when I visited day-care centers for our then-fifteen-month-old son, I was talking with a teacher about the food they served. She told me it was delivered by a national food distribution company and all of it was very healthy. The chicken nuggets, she assured me, were low-fat and all-white meat. To prove it, she handed over a binder detailing some of the commonly served foods. The list of substances used to make the chicken nuggets stretched across ten lines and included BHA, methylcellulose, the preservative propyl gallate, and soy protein, which is decidedly not white meat. It looked like someone's idea of the cheapest nugget they could make while still convincing people it was "healthy."

While a little BHA never killed anybody, a steady dose of quasi-edible food additives is of particular concern for children, whose small and still-developing bodies are more vulnerable to toxins than those of adults. Instead of being more wholesome, a lot of the food aimed at and favored by kids is less so than other foods. Much of it, for instance, is loaded up with artificial food colorings, which have been linked in studies, most notably one in 2007 that also included sodium benzoate and was commissioned by the British government, to hyperactivity and ADHD in kids. Products in the European Union that contain these additives now carry labels warning: "Consumption may have an adverse effect on activity and attention in children." The Feingold Association, a group founded in the seventies, has collected thousands of stories of children whose ADHD and other behavioral problems dramatically diminished when parents eliminated food colorings, artificial flavorings, artificial sweeteners and synthetic preservatives such as BHT and TBHQ from their diets.

But these remain anecdotal accounts and, in contradiction to the UK study, other research has shown that artificial colorings don't cause behavioral issues. The FDA has looked at this conflicting data and deemed there to be no clear cause for action. Until there is airtight, unqualified proof that Red #40 prompts children to run around madly in circles, the FDA is not going to take on the industry. It's just one example of an approach to regulation that happens to favor food companies, not parents.

Here's the way I look at it as someone raising two young boys: when I assess the risk of feeding our kids additives that are new to the human diet and are overseen by a porous, pseudo-regulatory system, I'd rather not take the chance—especially when there are so many alternatives. I'd rather feed them processed foods without chemical preservatives and ingredients that leave tongues Smurf-blue, or better yet, fresh, unprocessed foods that can be prepared at home.

What, after all, is the reward associated with all these risks? Certainly there's the ease of being able to buy foods that are readily available, low-priced, and enthusiastically welcomed by kids. Yet when I weigh this against Jude's and Luke's health and development, it isn't even a contest. Processed foods are designed to be irresistibly delicious and appealingly convenient, but the more you know about the story of food additives, the more hollow the appeal seems.

Biotech Baking

The food industry understands that increasing numbers of people are coming to the same conclusion. We represent a still-small percentage of the industry's customer base, but one meaningful enough that a trend called "clean label" has emerged. The exact parameters of what this means are still blurry. Some people use the Whole Foods list of seventy-eight "unacceptable food ingredients"—substances the company does not allow in products sold in its stores—as a guide. Others say clean-label processed food should be made with only ingredients people would have in their kitchen cupboards, though this is a high bar to clear. Generally speaking, the goal is to include no ingredients a health-conscious mom or dad would pause to wonder about while reading the side panel. It also means that ingredient statements can't be a mile long. Pillsbury's Simply frozen cookie dough contains fifteen mostly recognizable substances; Tyson 100% Natural chicken nuggets are made with eighteen; and Safeway's Open Nature Whole Grain White Bread has fifteen.

Making foods with fewer ingredients and fewer substances that might prompt hazmat alerts or are used to clean up oil spills is a step in the right direction. But by no means is clean label synonymous with healthy, low-tech, or minimally processed food. Look no further than Lay's regular potato chips, which contain a mere three ingredients—potatoes, vegetable oil, and salt—but aren't exactly healthy. The same goes for Tostitos Crispy Rounds, with corn, vegetable oil, and salt.

In Franklinton, a tiny town outside Raleigh, North Carolina, I got a more intimate look at the practical steps the industry is taking to try to move away from hazmat-suit additives. Like many towns across rural America, Franklinton has seen better days. It once thrived as a lumber stop on the New Hope Valley Railroad, but over the past twenty years, all three of the town's textile mills have either closed down or moved elsewhere. Such losses are reflected in the two-block strip of downtown, which looks one bad day away from becoming a ghost town. On the Monday afternoon I was there, Franklinton's sidewalks were barren. Dark and empty storefronts outnumbered the lighted ones, and just a handful of the angled parking spots contained cars, making a mockery of the two-hour parking limit signs punctuating the sidewalks.

Perhaps the only reason to come to Franklinton these days is to visit Novozymes, a Danish company that in the late 1970s bought acre upon acre of inexpensive land for its North American headquarters. The company, which last year had global revenues equivalent to $1.9 billion, is by far the biggest show in town, employing close to five hundred people. Most of these employees, though, are experienced scientists, engineers, and MBAs who have been recruited from elsewhere. Mostly, they reside not in Franklinton but in nearby Wake Forest, Raleigh, or Durham.

Novozymes's 1800-acre campus is about a ten-minute drive from Main Street. The road goes under the old train tracks and then along rolling hills and through a collection of small houses dotting the road. Some are cute, with neat shutters and chairs on their porches; many appear long abandoned. There's nothing remotely commercial or industrial about this landscape, and at some point I concluded that I'd missed a turn. That's when I saw a giant, shiny tanker truck barreling toward me and a smokestack thrusting through the trees.

Novozymes is in the enzymes business. Enzymes are a special type of protein made by every organism on the planet in order to facilitate chemical reactions. They break down food during digestion, move things around within cells, and help generate muscle contractions. Industrially produced enzymes are used in all sorts of products we've come to depend upon. They function as stain removers in laundry detergents, especially those billed as environmentally friendly; they turn corn into ethanol and switchgrass into biodiesel. Enzymes may have "stonewashed" the jeans in your closet, replacing petroleum-based chemicals. "Everything that's made from oil today can be made from sugar using enzymes," declared Paige Donnelly, a cheerful brunette who left Scottsdale, Arizona four years ago to work for Novozymes as its U.S. communications chief.

In the food industry, enzymes have been used for quite some time to turn cornstarch into high-fructose corn syrup. They also squeeze the last possible drop of juice from apples and get crackers to hold less water, thus making manufacturing easier. During the manufacture of vitamins C and B12, they rearrange molecules and catalyze reactions. Enzymes process soybean oil in a way that makes it trans-fat-free.

And then there's the job of cleaning up ingredient labels, or at least those of breads and other baked goods. Enzymes such as lipase, glucose oxidase, and xylanase can be used to replace those dirty dough conditioners and oxidizers—the mono- and diglycerides, sodium stearoyl lactylate, DATEM, and azodicarbonamide. Enzymes, it is thought, are natural, and they tend to pass the squint test with Whole Foods's soccer moms (my mom, however, remains skeptical; "What are they and how are they used?" she demands). Often, though, the word "enzymes" doesn't even appear on labels. Since they're inactivated by the heat of baking, enzymes are considered "processing aids" and don't have to be declared, though they remain (inactive) in the final loaf. Another category of enzymes, the so-called antistaling variety, is already in wide use. They help the vast majority of mass-produced loaves and buns stay soft and resilient for fourteen days, instead of four. Some breads note their use, others don't.

When Novozymes first started making enzymes in the 1940s, they did so in a way that was unequivocally natural—they isolated them from the pancreas of cattle. In the fifties, the company shifted toward the much more efficient mode of producing enzymes from bacteria. Today in Bagsvaerd, the town outside of Copenhagen where Novozymes is based, there's a space-age, walk-in freezer containing one of the world's largest stashes of preserved microorganisms—ten thousand different freeze-dried bacteria and fungi that can be used to make enzymes. Some of these, a very small number, have come from the wild—from teams of Novozymes scientists who collect soil and water samples from, as the company puts it, "the four corners of the Earth." The rest originate in a laboratory.

Nearly all the food enzymes produced in Franklinton have come from bacteria or fungi that have been genetically engineered to have very specific traits. Robots splice different microorganisms together in test tubes, giving the host organism genetic material that it couldn't have otherwise. With their new, highly selected DNA, these bacteria and fungi are placed inside two-story steel tanks along with a rich mixing broth. They ferment for anywhere from three days to two weeks, secreting enzymes as they go. Each tank will ultimately yield three or four truckloads of a sloshy brown liquid or multiple one-ton bags of dried tan powder. As we walked through the nearly spotless, fully automated plant, Adam Moore, Novozymes's blond, slender North American president, flipped open a porthole covering one of the tanks to show me these microscopic bugs in action. The mixture, which was being stirred by spinning blades to accelerate fermentation, looked like a river after a rainstorm, rushing thick and brown with mud.

While they're in this tightly controlled environment, Novozymes's genetically engineered bacteria excel at what they do. But once outside either test tubes or their closely monitored steel tanks, they wither and die. "They're designed for a high output of a specific enzyme, so once the conditions change, that's it. Most wild organisms are more robust," Moore said. I asked him whether this meant Novozymes's enzymes weren't natural, since the bacteria they're spawned from aren't found in nature and couldn't survive in it. He pointed out, quite rightly, that no legal definition exists for what's natural and what's not. The FDA has never defined the term, despite its unbridled use and a dozen court cases where food companies have been sued for using it. "Frankly, I don't see the difference between classic breeding and GM [genetic modification]," Moore added. "If anything, with GM you have a higher degree of safety because you're doing it in a controlled fashion." Novozymes has done varying amounts of safety testing on its food enzymes and has filed GRAS notifications with the FDA for all of them.

If you ask them, most Americans will say that genetically engineered ingredients don't belong in products advertising themselves as natural, yet it's an inclusion that happens all the time. In a 2011 report, the organic advocacy group Cornucopia found at least seven "natural" breakfast cereals containing substantial amounts of GM ingredients, including the corn in Whole Foods's 365 Corn Flakes. The GM classification of enzymes is a lot more subtle, because while the parent bacteria are genetically modified, the enzymes themselves aren't. Such GM-generated enzymes can't be used in certified organic products but they're routinely included in so-called natural ones.

Standing outside Novozymes's baking lab in front of a wall decorated with vintage bread ads, Todd Forman, one of Novozymes's baking scientists, showed me what the company's clean-label enzymes can do for bread. He clutched a bag of Whole Foods sliced organic white bread that he'd recently purchased and urged me to examine it. "Look how small it is," he commented. "And dense." He hoisted it into my hands. He was right—not the least bit fluffy. Because the bread hadn't been baked with dough conditioners and oxidizers or similarly functioning enzymes, it had what baking scientists call inferior "loaf volume." It was less air, more bread—thus more expensive to produce.

Forman explained that he could, in theory, take the Whole Foods doorstop into his lab and remake it with high-speed mixing machines and enzymes, boosting its size by more than one third. He does this sort of thing every day, helping customers bake bread more economically and with a measurably fluffier texture. I watched as a dedicated machine plunged a small, translucent plastic disc into the center of a bread slice, calibrating its "springiness," "cohesiveness," and "adhesiveness."

There are more old-fashioned testing methods, too. "Sometimes I think I eat the equivalent of a whole loaf in fifteen minutes," Forman sighed. He was referring to the way his laboratory experiments often have to be personally sampled. One of the best ways to assess bread's qualities is still to taste it. As we talked, I could see a dozen pans of perfectly golden loaves emerging from the oven. They smelled like heaven. Baked with various Novozymes products, they were also 100 percent natural, at least relatively speaking.

7

The Joy of Soy

A man is only as old as his arteries.

—Thomas Sydenham, seventeenth-century English physician

Back in the mid-eighties, when I was in high school, there was one fact about eating that everyone knew for sure. You didn't have to be well-read or even particularly interested in health to know it. The message was everywhere, loud and clear: Fat was bad. Eating any significant quantity of it was an affront to both your arteries and your waistline. Egg yolks, bacon, and avocados were banished. When I ate pizza, I always kept a stash of fat-blotting napkins nearby. Supermarket shelves were stocked with ample supplies of formerly fatty foods—low-fat ice cream, cookies, muffins, cakes, peanut butter, hot dogs, cheese, and coffee creamer. I remember reading at one point that even oatmeal had fat lurking within its cocoon of carbs. Should I stop eating oatmeal, I wondered in full seriousness?

Thankfully the misguided mania over dietary fat has passed. There are still prominent vestiges of antifat orthodoxy out there, but most scientists now realize that this nutrient is crucial from a biological point of view. It also turns out that the types of fat we eat matter a whole lot, and not just to our heart and arteries. Fats are necessary for cell membranes; they make up 60 percent of our brains, and are critical for the production of neurotransmitters. Fats, essentially, are us.

Over the past three-quarters of a century, our fat consumption has shifted profoundly. Not only are we eating it in greater quantities today (double the levels in 1909), we're also getting a big chunk—some 10 percent of daily calories—from a source that didn't exist a hundred years ago, a type of oil only made edible and commercially possible through the use of modern chemical technology and machinery.

Soybean Fields Forever

Driving along the flat band of interstate from St. Louis, Missouri to Decatur, Illinois, the fields of stately corn stalks periodically give way to stretches of squat, stubbly plants. I wasn't sure what they were at first, but after stopping along the highway a few times to get a better look, I realized that pretty much anything that wasn't corn was soybeans.

An unassuming crop, they grow in pods attached to leafy stalks, with the leaves about the size and shape of basil. Some 55 percent of the entire state of Illinois is covered with corn and soybeans. At one point I was at a gas station where soybeans crawled right up to the edge of the parking lot. Later, at a KFC just off the highway, corn almost seemed poised to enter the restaurant, as if it had been possessed with a premonition about its ultimate destiny.

While corn has a long history of cultivation in the United States, predating Columbus's arrival, soybeans haven't been here for very long. The crop first found its way to our shores in the early 1700s from China, where it had been grown since 2,700 BCE, mostly for fermented foods like soy sauce. It wasn't planted in any significant quantity here in the United States for another two hundred years, and then farmers grew it not for food but as a source of fertilizer. Because soybeans can take nitrogen from the atmosphere and trap it in the soil, they were grown following the corn harvest and then plowed under. When it was discovered that this compost crop had a lot of protein, farmers started growing soybeans to feed cattle and pigs—species that had never before eaten soy as part of their natural diet but tended to grow much faster when they did.

Today, the 74 million acres of soybeans planted throughout the American Midwest and South continue to be harvested for animal feed, though they're now equally important as a source of oil, most of which is consumed by humans in the form of processed food. Of the 14 billion pounds of soybean oil produced in 2011 for food, only a small fraction is used at home for cooking with Wesson or some other brand of "vegetable oil" (it's rarely labeled soybean).

Most forms of processed food need some fat, and for the past six decades, soybean oil has been the cheapest form available, thanks to the soybean's ease of cultivation and government farm subsidies totaling $1.5 billion a year. At supermarkets, soybean oil is in salad dressings, cookies, crackers, chips, chicken products, frozen dinners, peanut butter, and imitation pizza "cheese." At fast-food restaurants, it's in the sauces, buns, tortillas, and the fryers. Other vegetable oils like canola, corn, sunflower, and palm are also widely distributed within the food supply, but soybean has the biggest share, commanding 62 percent of the market for added oils (a diminishment from its 80 percent share in 2004).

It should come as no great surprise that this one ingredient has had far-reaching nutritional consequences. Before recent times, humans didn't consume vegetable oils in large quantities. Generations of Asians have eaten soy made from whole beans in the form of soy sauce, tempeh, natto, miso, tamari, and tofu—but the oil was not widely consumed. Historically, the fat in our diets coming from seeds, nuts, and other plants was bound in its original packaging. The only isolated vegetable oil eaten in any significant quantity was olive oil. Joe Hibbeln, acting chief of nutritional neurosciences at a research division of the National Institutes of Health, calls the rise of soybean oil "the single greatest, most rapid dietary change in the history of *Homo sapiens*." It sounds like an overblown claim until you take a step back and look at how these humble beans have crept into every corner of our diet.

In 1909, most of the fat we cooked with or added to food was in the form of lard (pork fat) and butter. In total, 82 percent of all of our fat came from animals. When Oreos were introduced in 1912, they were made with an entirely lard-based filling. That was what people ate. By 1950, cheaper soybean oil had dislodged butter as our top source of added fat, and by the mid-fifties, lard, too, was in rapid retreat.

Now the animal-vegetable divide is flipped: 44 percent of our total fat intake is animal and 56 percent vegetable, predominantly coming not from plants themselves but from extracted, refined industrial oils.

Explosion-Proof Factories

Decatur, smack dab in the middle of Illinois, is synonymous with soybeans. They first went into mass production here in 1922 when a starch salesman named A. E. Staley set up an experimental plant for making edible soybean oil. At first there wasn't much of a market for this. Soybean oil had a terrible taste and smell, with aromas that have been described variously as beany, grassy, and painty. In its early days, it was used mostly for making things such as paint, glycerin, linoleum, enamel, waterproof goods, and printing ink. By the time a small linseed oil producer named Archer Daniels Midland, now known simply as ADM, opened a similar Decatur factory in 1939, some—but certainly not all—of the oil's smell and taste problems had been solved. Both companies rode the wave of soybean expansion. Soon there was so much of the crop migrating through the town that the USDA started referring to soybean prices as "crude Decatur."

Today, the soybean still looms large over the town. The WSOY radio station calls itself "The Voice of Decatur," although the town's political persuasions are more mixed than the conservative AM talk-radio patter implies. You can open a Simply Soy checking account at the Soy Capital Bank and Trust or, for $45, stay at the Soy City Motel, located in the shadow of that original A. E. Staley plant, which is still humming today on a vastly larger scale. Now owned by the British food ingredient company Tate & Lyle, the factory processes corn, making high-fructose corn syrup and various starches. Down the road is ADM's factory and global headquarters, also much expanded. These plants fill the town with the constant whir of heavy machinery and an omnipresent smell reminiscent of burnt, rotted corn.

With 4,000 employees at ADM and 822 at Tate & Lyle, the companies also give Decatur, a struggling town of 76,000, a good deal of its economic backbone. Mayor Mike McElroy remembers that in the sixties the nuns at his Catholic school would chide students for complaining about the late summertime odor that smothered the town, apparently worse in those days. "That's the smell of money," the nuns would say wisely.

There's not as much soybean and corn money in Decatur as there used to be, in part because all the various associated operations moved elsewhere as the food industry consolidated. But there's still quite a lot of soybean oil here, thanks to ADM, a global behemoth with annual sales of $88 billion. About the size of a college campus, ADM's complex sprawls in all directions. In total, half a dozen different factories operate on the site, including ones that process corn. Some of these structures were built recently, while others were expanded with the haphazard attachment of a modern-looking structure onto an aging one. At the company's outset, no one could have anticipated the way its markets would grow.

The day I was there, a steady thunder of semis and tanker trucks lumbered along the main road leading to the compound. Rail cars, too, were coming and going, skidding along the wreath of tracks encircling the campus. I counted twenty-nine dormant hydrochloric acid tankers in one rail yard. At night, the place was lit up like Times Square, with steam still pouring out of smokestacks.

I wasn't allowed in for a tour of ADM's soybean oil plant; the PR rep explained the company doesn't like to do plant visits. Nor was I offered a visit to any of the facilities operated by the three other companies that together with ADM produce the vast majority of the soybean oil going into our food. Bunge didn't return calls or respond to e-mails. Cargill said they could perhaps let me see one of their soybean facilities, and then stopped responding. Ag Processing, a farmer-owned co-op, said they don't do tours for liability reasons, but sent a video explaining how soybean process-ing works. ADM was helpful in answering some questions.

Despite the impression this sort of reticence might create, there's not much reason for secrecy here. What happens inside ADM's Decatur factory is more or less the same thing that occurs inside all the others, and it's been happening this way since the sixties with only with minor modifications. The lengthy, intricate process illustrates why soybean oil is not a "simple ingredient," as Frito-Lay calls it on Tostitos packages, or "100% natural," as Wes-son refers to it, but a complex, high-tech product. Soybeans are hard, sturdy pebbles and are not easily transformed into flowing vats of liquid. Unlike olives, which ooze oil, and cream, which if left alone separates easily from milk, quite a lot has to be done to pry oil out of soybeans and make it suitable for processed foods. In fact, some ninety years after its introduction, we're still searching for new technologies for what has been a troublesome oil from the beginning.

Whole soybeans are dispatched into the Decatur plant from a row of nearby grain silos fed by the daily arrival of rail cars. Each soybean is dried, dehulled, cracked, flattened into flakes, and then sent for de-oiling. For the past five decades, this has been done with a chemical solvent that requires soybean oil factories be explosion-proof. Its main component is hexane, which comes from the same fraction of crude oil that's used to make gasoline. Hexane is clas-sified by the Occupational Safety and Health Administration as a neurotoxin. After just ten to fifteen minutes of exposure, vertigo, headaches, nausea, and eye and upper respiratory tract irritations can develop. Repeated exposure at high levels can lead to muscle weakness, numbness, and neurological disorders. In early 2012, Chinese workers at a factory making iPhones claimed to have suf-fered health problems after being forced to use hexane to clean the screens. One worker said hexane left him with such nerve damage and hypersensitivity to cold that he wears down-insulated clothing indoors.

The Environmental Protection Agency (EPA) also keeps tabs on hexane because, as a neurotoxin, it has the potential to cause permanent nerve damage in humans. According to the EPA's database of hazardous air pollutants, the vegetable oil industry is the largest emitter of hexane, with ADM's Decatur facility topping the list.

Fortunately, plant workers in Decatur are not subjected to much in the way of hexane emissions beyond what's present in the external air. Inside factories, the chemical is housed safely within pipes and machinery. Yet precautions are taken. Due to its flammability, all switches and motors on machinery need to be enclosed, and signs must warn workers about the proximity of flammable materials and the possibility, though rare, of explosions. The last major one was at an Ag Processing soy biodiesel facility outside Sioux City, Iowa. In 2003, during a routine shutdown, a malfunction in the plant's ventilation system caused gasses from vats of hexane to accumulate and then ignite, killing two workers and injuring six.

The soybean industry's widespread adoption of hexane in the sixties helped boost oil production. The previous method of hydraulic expeller pressing—something still in use today for organic, where the use of hexane is prohibited—squeezed out only 70 percent of the oil while requiring more energy than solvent processing. Hexane enables 99 percent extraction. Over the years, the industry has researched alternatives, but none have matched hexane's effectiveness and efficiency. Other vegetable oils such as canola, corn, sunflower, and cottonseed are similarly processed with the chemical.

After it's leached the oil out, hexane is separated from the mixture by vaporization in a vacuum. The hexane is then cooled, condensed back into liquid, and reused. ADM and other soybean processors say that any traces of hexane remaining in the oil or in the solids from which the oil is removed are just that—traces. Our biggest exposure to hexane is likely coming not from food but from the gasoline fumes we breathe in at gas stations, since motor fuel contains small amounts of the chemical. But how much we're exposed to in our food is hard to say exactly. The FDA doesn't set a maximum residue level or require companies to do any testing. Wilmot Wijeratne, director of food technology at Harvest Innovations, one of the few companies selling non-hexane-processed soy ingredients, told me that hexane residues of 50 parts per million in soybean oil aren't uncommon. "As a scientist I don't think this is a problem, but there are some 60 million Americans that are sensitive to chemicals and it's all about the perception of the consumer," he said.

Once pried from the bean, soybean oil is the color of weak coffee and nowhere near ready for frying nuggets or drizzling onto a salad. If you tasted it at this juncture, it would be "a bit like eating grass," said Duncan Guy, a technical sales rep for Bunge. To fix this, the oil is piped to an adjacent facility at ADM's complex, where it's treated with two other chemicals—sodium hydroxide and phosphoric acid—to remove any lingering soy solids, as well as some of the dark color, pesticides, and other various impurities. Then it's bleached using a clay filter and treated with hydrochloric acid, the stuff I saw sitting in the rail tankers. This process removes all color pigments, including the red-hued compound beta-carotene, the substance our bodies convert to vitamin A. If not for the bleaching, soybean oil would be healthier, but also a bright reddish-orange color.

The next step, deodorizing—which happens in a vacuum at 500°F—saps the oil of every trace of odor and flavor and also of its vitamin E and phytosterols, substances thought to block the absorption of cholesterol. Like lecithin, a gummy substance removed from soybean oil before bleaching, these healthy compounds are marketed separately, which is more financially advantageous than adding them back into the oil. ADM sells twelve different vitamin E products, sixteen lecithins, and a brand of plant sterols called CardioAid that you can find in yogurts. Although they don't use hexane, natural and organic vegetable oils also go through similar bleaching and deodorizing steps.

Vitamin E—a potent, natural antioxidant—is usually replaced with the preservative TBHQ or a slightly more natural one like citric acid. If the oil is destined for deep fryers, it will also have the silicone-based chemical dimethylpolysiloxane added to keep the oil from getting frothy after countless rounds of frying. (Lots of foam means frying vats can't be filled to capacity.) Then, at long last, after multiple rounds of heating, chemical treating, centrifuging, filtering, washing, and vacuuming, the soybean oil—clear, tasteless, odorless, nearly nutrient-free—is edible, ready to be pumped into eight-thousand-gallon tanker trucks and shipped off to customers across the food industry.

Or in theory it's ready. For many decades there was something else that needed to be done to it first.

Melts in Your Mouth

By nature, soybean oil is an unstable polyunsaturated fat. Unlike the saturated animal fats it helped replace, it spritzes a particularly hard-to-clean, waxy grease when heated. At the IFT show in New Orleans, a soap manufacturer received a special award for inventing a product that cleaned up the stubborn gunk that splattering soybean oil leaves in its wake. Also when heated, soybean oil breaks down, or oxidizes, which over time can result in rancidity and stale flavors. Since most food processing involves significant levels of heat, there's always the risk that the oil will infect whatever product it's in with a pungent painty or wet-cardboard odor—unless it's further processed.

Until recently, this meant partial hydrogenation, a relatively basic technology that was developed at the turn of the twentieth century and made soybean oil a lot more durable and less prone to splatter and oxidation. Partial hydrogenation alters the oil's molecular structure so that what was liquid becomes solid—less like oil, more like butter. "The beauty of partially hydrogenated oil was that it melted in your mouth. Literally," Mark Jackson, a sales rep from Bunge, a commodity processing company, told me, as we sat on one of the company's couches at IFT. "Its melting point is the temperature of your mouth."

Partial hydrogenation remained an industry mainstay until the late nineties, when the rap sheet against trans fats, which are inadvertently formed by the process, became impossible to ignore. A few prescient scientists had raised alarms about partial hydrogenation in the seventies, but no one wanted to believe that this technology—which entails putting a metal such as nickel into heated vegetable oil and bubbling hydrogen gas through it, forcing its fatty acid molecules to accept a few extra hydrogen atoms—could be causing such health problems. The news that trans fats clogged arteries and interfered with cellular activity was a huge blow. Starting in 2005, food manufacturers vowed to remove these dangerous by-products from their foods, sending processors like ADM and Bunge back to square one with soybean oil—waxy splatter; rancidity; an ingredient that might start smelling like wheatgrass; liquid when you needed a solid. "The functionality of partial hydrogenation just can't be fully replaced," Jackson lamented. Bunge and others have tried to address this by blending soybean with other oils, diluting its negative effects while maintaining its cost advantages.

Yet now that trans fats have been nearly banished from the food supply, another lurking problem has come to light. It's something that the practice of partial hydrogenation partly covered up and that someone only recently thought to look for.

The first time A. Saari Csallany heard of something called hydroxynonenals, it was the summer of 1991. She had arrived early one morning at her office to catch up on some scientific journals piling up on her desk. Not long after settling in with her coffee, she started thumbing through a publication called *Lipids* that contained an article by an Austrian biochemist named Hermann Esterbauer. In it, Esterbauer summarized his pioneering work on how breakdown substances called hydroxynonenals are formed in the body. Various kinds of internal, oxidative stresses, he wrote, cause the fats located in our cells, specifically the polyunsaturated kind, to dismantle and form various unhealthy compounds, some of which are particularly reactive and serve as "toxic messengers" that cause free-radical damage inside the body.

For a scientist studying the chemistry of fats, this was riveting. After studying chemistry in her native Hungary, Csallany (pronounced *Challani*) had come first to the University of Illinois, then to the University of Minnesota in 1973 to be an associate professor in the department of food science and nutrition. For several years, she studied vitamin E—"the best biological antioxidant," she says—and then became interested in the opposite principle, the mechanism by which fatty acids oxidize, and what happens when they do. Hydroxynonenals, which are also referred to as toxic aldehydes, were a new type of compound, and she wanted to know more about them. Csallany, who speaks with a thick Hungarian accent (she answers her phone with a curt "Csallany"), spent a month at Esterbauer's lab at the University of Graz, observing his methods and soaking up his knowledge.

Back in Minnesota, she began to wonder if toxic aldehydes were formed in other types of fats, namely the ones we consume in our diets. Soybean oil seemed an obvious place to start. In her lab, she heated the oil—a basic, nonhydrogenated variety from a Cargill plant in nearby Wayzata—to the standard frying temperature of 365°F, and analyzed it using Esterbauer's methods. What she found was that one particularly malignant type of toxic aldehyde called HNE, which wasn't present in the unheated oil, started forming after thirty minutes and increased for about six hours, at which point it leveled off and even decreased slightly. Her summary of this experiment, published in the *Journal of the American Oil Chemists' Society* in 2002, didn't attract much attention. To Csallany, though, the implications for public health were immediately obvious. She had identified what was causing HNE to form—the heating of a polyunsaturated fatty acid called linoleic acid found in large quantities in soybean oil and other commonly used vegetable fats like corn and cottonseed oils. (Partial hydrogenation removes some of the linoleic acid, so presumably the HNE levels would be lower, though Csallany says she hasn't tested partially hydrogenated oil.)

Csallany then wanted to figure out if this troubling compound was migrating from the hot frying oil into food. The unfortunate answer was yes. She fried pieces of potato in soybean oil, and after two hours the french fries had concentrations similar to that of the oil. "Excessive consumption of fried foods could be a health concern," she noted dryly when the study was published in the same journal in 2004, just as McDonald's and every other fast-food chain was trying to figure out how to ditch trans fats.

Were HNEs present in the real world outside the lab? In 2010, Csallany sent a team of graduate students out to six different major fast-food restaurants in Minneapolis—she doesn't want to name them—to buy french fries every two hours for a period of twenty-four hours. After each purchase, the students shuttled the fries back to the lab in St. Paul and then headed back out. When all the tests were done, the results showed HNE concentrations ranging from 7 to 32 parts per million, with an average of 13. At this point, most fast-food restaurants had replaced their partially hydrogenated soybean oil with a mixture of several oils, most of them containing significant concentrations of linoleic acid.

The issue of toxic aldehydes in soybean oil has received no mainstream attention, but ADM, Cargill, and Bunge are all aware of their formation and potential health implications. According to the work of dozens of scientists now researching the subject, toxic aldehydes are so reactive that they can interfere with both enzyme and hormone production as well as with the basic act of protein synthesis. They're thought to be possible causal agents for diseases and conditions ranging from Alzheimer's, Parkinson's, atherosclerosis, cancer, and chronic inflammation. Nearly all these scientists, though, are studying the formation of these bad actors inside the body. Csallany remains one of only a handful focused on toxic aldehydes in the diet. Thus there's still work to be done identifying the specific health implications.

I asked Csallany whether her work had altered her own eating habits. Does she ever eat fried foods? "No, no, no. I'm not frying anything in any of these oils," she said. "I definitely don't eat french fries anymore."

Allan Butterfield, a chemistry professor at the University of

Kentucky and someone who's linked HNEs to Alzheimer's, told me he doesn't think oxidized vegetable oils would be a major concern in Alzheimer's disease. More likely, he said via e-mail, they would cause problems in our arteries, muscle tissues, and organs like the liver—the periphery as opposed to the nervous system.

The way ADM sees it, HNEs aren't a problem because they're present in oils at such low concentrations. Several years ago, the company had researchers at the University of Illinois replicate Csallany's tests on frying oil. They used a blend of corn and soybean oil and found HNEs at a concentration of just several parts per million, versus Csallany's range of up to 32 parts per million. "There are a large number of toxicologists that have looked at these compounds and identified that they're clearly undesirable," acknowledged William Artz, an associate professor of food processing and one of the researchers involved in the study. "But at these low concentrations I think the body can deal with it," he said.

So consuming less partially hydrogenated oil has ironically led us to consume more toxic aldehydes, which may actually be worse for us. Nearly a decade into the post-trans-fat era, no clear-cut winner for replacing partially hydrogenated soybean oil as processed food's favorite fat has been found. No alternative possesses its wide utility. Corn oil retains a corny taste; canola can only be grown in northern climates, limiting its production; sunflower and safflower are too expensive; and butter is way too expensive, as is coconut oil. Palm oil is useful, but manufacturers and fast-food companies are reluctant to use too much because it's a saturated fat, something they're still trying to take out of their products; also, it's not a product that influential U.S. farm interests can get behind. Palm comes mostly from Malaysia and Indonesia, along with a trail of messy environmental PR. Increased demand for the oil in recent years has caused widespread deforestation across large areas of both countries, destroying all sorts of fragile habitats and threatening one of our closest evolutionary relatives, the orangutan, with extinction.

Soybean oil is also deeply intertwined with the fate of billions of acres of Midwest farmland and with the business models of powerful companies like ADM, Cargill, Bunge, Monsanto, and DuPont. "Yeah, we're married to soy, we're a soy company," said Jackson, when I asked him if Bunge still needed to find ways to make soybean oil viable. Along with its competitors, Bunge is hard at work on new high-tech answers.

One solution is already available. A thirty-year-old technology called interesterification plays a game of molecular musical chairs with fatty-acid molecules to make them more saturated. Its proponents say it does so without the negative health effects of partial hydrogenation. ADM has worked with Novozymes to create an enzyme that interesterifies soybean oil. Bunge also utilizes enzymes. But it's not as cheap as hydrogenation, and there are some potential red flags. K. C. Hayes, a biologist at Brandeis University, has done studies showing that interesterified oils can raise blood sugar and impair insulin secretion, in addition to negatively affecting cholesterol metabolism. "When you randomize the fatty-acid profile of fats from what nature intended, you can cause problems," Hayes said. "I don't think we know the full answer yet. I'm just raising the question."

Monsanto and DuPont are taking the fight upstream, to the seeds. Both companies are rejiggering DNA to make a soybean's fatty-acid profile more like that of an olive. Still under development, this new generation of so-called high-oleic (and low-linoleic) genetically modified soybeans will yield oil that's more stable. DuPont expects farmers to begin planting them in 2013 and Monsanto the following year.

Something Fishy

The science of dietary fats has got to be one of the most eye-glazing and head-spinning things in all of human nutrition. The whole business is laden with contradictions and changing viewpoints. First all fats were bad, then it was just saturated fats. The poly-unsaturated varieties like corn oil and soybean oil (PUFAs) and monounsaturated fats like canola oil (MUFAs) were laid out like an all-you-can-eat buffet. Companies championed their use. "The moofas and poofas reduce your bad cholesterol, so not only did we take out the bad, we added in something good," Rocco Papa-lia, then Frito-Lay's senior vice president for research and develop-ment, told me in 2005, not long after the company replaced its partially hydrogenated vegetable oils with a mix of unhydroge-nated ones.

Today, if you listen closely, you'll hear talk that would have been unimaginable a decade ago—suggestions that some saturated fats, such as stearic acid, which is found in chocolate, among other things, might be beneficial. In 2010, the government's Dietary Guidelines stated that "stearic acid has different metabolic effects than other SFAs [saturated fatty acids] and does not raise blood cholesterol." Some nutritionists have also begun to acknowledge that coconut oil, that most saturated of fats, might also have ben-efits.

The more you look at it, the better those saturated fats—the ones that don't need to be chemically extracted or molecularly rearranged—start to look. As scientists like Tom Sanders at King's College in London will tell you, saturated fats are nowhere near as hostile as we once thought. In fact, they appear to have little to no correlation with heart disease. If anything, the highly processed vegetable oils we've been consuming by the boatload for at least three decades may be much bigger culprits.

The biggest reason for considering this actually has nothing to do with soybean oil on its own. Rather, it's the monumental quantities we're ingesting that distort our bodies' critical ratio of omega 6 and omega 3 fats. The ideal proportion is somewhere between one and three omega 6 fats to every one omega 3. That's roughly what it's been in most cultures throughout human history. Today the ratio in the American diet is about ten to one, and our cells are flooded with omega 6s. This dietary imbalance is thought to have a number of far-reaching implications for human health. Many in the medical community think it is a contributor to heart disease, many forms of cancer, depression, and various other diseases that stem from inflammation. Soybeans in our diet are the primary reason for this dangerous imbalance.

Don't get me wrong: I love the taste of soy milk in lattes, and there's nothing wrong with a little (unhydrogenated, unheated) soybean oil, though other oils offer much more flavor. Tofu with the right seasonings can be delicious, and tempeh is incredibly nutritious, in part because it contains healthy bacteria. But with soybean oil, which is about 55 percent omega 6 (35 percent when it's partially hydrogenated) flooding the market, the powerful soybean industry has made our health the victim of its success.

We're also getting more omega 6s thanks to the fact that chicken, pigs, turkeys, and cattle now consume diets high in soy meal and corn, which also has off-the-chart levels of omega 6. The animals' traditional pre-twentieth-century diets of grass and other wild foods supplied them with omega 3s. In a 2011 study, Joe Hibbeln of the National Institutes of Health (NIH) found linoleic acid levels (omega 6) in modern ground beef, pork, bacon, chicken, beef tallow, and lard higher than in animals raised according to practices used in 1909—methods that Hibbeln asked Joel Salatin of Polyface Farms in Virginia to replicate for the study. They ranged from 12 percent and 40 percent higher for chicken and pork, to 200 percent and 300 percent higher for bacon and ground beef. Modern omega 3 concentrations were higher too, but not enough to offset the leaps in omega 6.

Soy wasn't always a dietary villain. As overconsumed as it is today, it has served worthwhile purposes in the past. As an ingredient consistently cheaper than animal fats, it's helped lower the cost of foods that contain it. And during World War II, soybean oil provided a steady, cheap source of calories for both Americans at home and in the foods sent over to Europe to feed U.S. and Allied troops.

Today, it's a different war, one to get America's omega balance back in line. The obvious solution here would be a recommendation that Americans scale back on omega 6 fatty acids, such as soybean and corn oil. But that's not the message getting delivered. Instead, many health experts are urging us to eat more fish, particularly fatty fish like mackerel, trout, tuna, sardines, and salmon, which are excellent sources of the so-called important long-chain omega 3s, also known as DHA and EPA. These fats are extremely beneficial, essential for brain development in babies and children, and helpful in lowering triglycerides and reducing inflammation. The American Heart Association, the biggest, most mainstream medical group overseeing all things to do with heart disease, urges two servings of fatty fish a week.

No doubt eating fish is healthy, but if the goal is a healthy omega ratio, we're not going to get there by doubling down on seafood or popping fish oil supplements. In this case, as in so many others in human biology, optimal health isn't so much about absolute quantities as it is about balance. Just as the ratio of your HDL to LDL cholesterol is much more revealing than total cholesterol, and elevated sodium consumption can sometimes be offset by adequate intakes of potassium, so omega 3 consumption, too, needs to be in proportion with omega 6. According to Hibbeln, in order to get this balance right at our current rates of omega 6 consumption, we would need to ingest either fourteen 500-milligram capsules of fish oil a day, or between six and ten ounces of salmon—levels that would stymie even the most avid seafood lover, not to mention deplete the oceans. The American Heart Association, the committee responsible for the government's Dietary Guidelines, and other health authorities don't dare say it, but the only way to right the ship is to trim back on omega 6–rich vegetable oils, which invariably means eating less processed food.

Doing this will allow our bodies to access another useful stash of omega 3s—the plant-based, short-chain ones contained in walnuts, flax seeds, green vegetables (yes, spinach and chard have some fat), and even soybeans. Our bodies can convert these fats into the long-form DHA and EPA found primarily in fish, but they can't do it when too many omega 6 fatty acids are present, because both types compete for the same metabolic pathways in the body. (This offers further evidence that soybean oil is ill-suited for mass consumption; it has seven omega 6s for every one omega 3.) Hibbeln equated it to thirty football players and one cheerleader all trying to enter through a single doorway. "That cheerleader's not getting in," he said, in a phone conversation. "There's only so much space in the body, and you have to back off the omega 6's substantially, from 8 to 12 percent of calories in the diet now to more like 2 percent."

Or maybe Monsanto will fix the whole thing for us—technology coming to clean up what it unleashed in the first place. At the same time Monsanto scientists are retooling soybeans' DNA to reduce its unstable fatty acids (which include both the omega 3s and omega 6s), they're also working on developing Soymega seeds, which are high in a unique type of omega 3 called stearidonic acid. This type of fatty acid is found not in foods but plant products such as evening primrose oil. To hear Rick Wilkes, Monsanto's director of food applications, tell it, stearidonic acid is more beneficial than the plant-based omega 3s we're eating now because it's more easily converted by the body into the long-chain forms. "Consumers will have the choice of increasing long chain omega 3s in their diet without having to eat things they turn their nose up at, like mackerel and herring," Wilkes said.

What does the future hold for the shape-shifting soybean? The answer will no doubt be dreamed up by an enterprising food scientist. On an IFT Web site, a video called "What Is Food Science, Anyway?," intended to make the profession appealing to young people, features half a dozen scientists talking about why they like their jobs. Mitchell Duffy, who at the time the video was made was a bakery and cereal specialist at the flavor company David Michael & Co, declared that "everybody wants the next new thing." His big innovation, he said, was "Jell-O pudding in a cup." Eric Shinsato at Corn Products International said gleefully, "They pay you to play with food." Amanda LaCoste works as a senior food technologist at Lightlife Foods, a ConAgra brand of vegetarian products. "You take a fresh soybean, and I have to make that look and taste like a hotdog. How the heck do you do that?" she said with a big grin. "That's what I do."

8

Extended Meat

*Inhabitants of underdeveloped nations and victims of
natural disasters are the only people who have ever
been happy to see soybeans.*

—Fran Lebowitz

In October 1959, just a few days shy of Halloween, a dozen corporate executives and factory workers gathered at an industrial site on the outskirts of Chicago to celebrate what many of them perceived to be a historic event. It was the opening ceremony for a factory manufacturing edible soy protein. In a triumphant speech, Harold McMillen, chairman of the board of Central Soya, the company that built the plant, dedicated it to "the world's growing population, for whom protein provides the building blocks of good nutrition and health." Another executive compared the incipient production of soy protein to the world-changing Sputnik launch two years earlier. He predicted that "no metallic satellite in outer space will be able to match in terms of human happiness and well-being the contribution of this, the protein satellite." Though surely a far cry from today's sophisticated offerings, the world's first meatless hot dogs were served.

Prior to this, nobody had ever thought to include soy protein and Sputnik in the same sentence. Nor had they ever consumed something called soy protein, much less in the form of a hot dog. Soybean oil gave Americans their first taste of what were once called "Manchurian beans," but it wasn't long before the burgeoning soybean industry figured out that this crop had more parts to be poked, prodded, extracted, and ultimately turned into dinner. Like soybean oil, though, it was going to take quite a lot of scientific brain power to make what remained of soybeans after de-oiling—the soybean meal—edible.

Initially soy meal went primarily into industrial products such as paper coatings and fire-fighting foam, a product responsible for saving the lives of many Navy sailors during World War II. Henry Ford famously made car parts and eventually an entire vehicle whose body had been constructed from plastic produced from soybeans. He also wore a wool "soybean suit" on special occasions. By the late forties, soy wool was consigned to the dustbin of sartorial history, and most soy meal was going to animal feed. Soybean crops are a useful generator of growth-boosting protein. By weight, a soybean is 35 to 40 percent protein.

But as for human consumption? In its raw form, soybean meal doesn't taste very good and contains a compound that blocks the absorption of essential minerals. But with greater and greater quantities of soy meal being generated every year, scientists at Central Soya, seeing the prospect of a food ingredient far more lucrative than animal feed, set out in the forties to tackle these obstacles. Within a decade, they'd devised ways to process the beans, including steam-heating and toasting the soy meal, in order to minimize the grassy, beany flavors, as well as get rid of the nutrient-blocking phytic acid.

And then it was just a matter of marketing the new ingredient to food processors. Candy manufacturers were among the first takers. They used it to prevent their products from sticking to the wrappers. Soon after, Central Soya and other manufacturers showed dessert makers how soy protein could taste and act remarkably like whipped cream. General Mills started manufacturing a series of puddings and nondairy desserts with it. The company became so enamored with the ingredient that in 1969 it started its own soy protein production, going on to become the first manufacturer of imitation bacon bits. Makers of actual meat also welcomed soy protein, using it as a filler or extender in their bologna, salami, ham, and turkey slices. The meat manufacturer Swift & Co started its own production of this remarkable ingredient.

In 1969, a *New York Times* article explained the marvel of it:

> The potential of the food is virtually unlimited. . . . Products can be tailored into any desired framework. These include vegetarian, Kosher, polyunsaturated fat, high or low in carbohydrates or animal or vegetable fat, zero cholesterol, with or without vitamins and minerals, and precisely controlled calorie content. They can be refrigerated, frozen, canned, or dried.

Soy protein was, in other words, the perfectly malleable medium for food scientists—at once the blank canvas and the paint.

To one degree or another, it still is. You can find soy protein alongside chicken in KFC's Country Fried Steak and Original Recipe Filet, even though no soy protein existed when the Colonel first concocted his famous recipes. Soy protein also helps out Subway's chicken breast strips and meatballs, and it's in Hardee's chili and its steak. At the supermarket, soy protein is found in bars, breakfast cereals, protein drinks, frozen entrees, soy-based infant formula, fat-reduced peanut butter, tuna fish, veggie burgers, and, of course, veggie hot dogs. And because of the way it can mimic meat, soy protein forms the backbone of the vegetarian products industry. Anything that's designed to look like meat but isn't is probably constructed from soy protein.

Soy Swapping

What's left of Central Soya is now Solae, by far the world's largest maker of soy protein. Last year it sold roughly $1.3 billion worth of the stuff (along with soy fiber, lecithin, and soy polymers for industrial applications) in eighty countries, some $410 million of it in North America. Formed a decade ago from the merger of long-standing business units belonging to the chemical and seed company DuPont and the soybean processor Bunge, the company's sole purpose is to do what its predecessor Central Soya did in the sixties—find new ways for soybeans to be used in foods. More so than perhaps any other company, its future is wedded to the soybean's fate as a food ingredient, which is why Solae, now fully owned by DuPont, is working with DuPont's competitor Monsanto on that omega 6–rich soybean oil.

Located along the western edge of St. Louis, Missouri, Solae's headquarters are housed in a sleek U-shaped office building that looks like a stainless-steel and glass castle amid the abandoned warehouses and factories. On the sweltering July day I was there, a fenced-in patch of well-groomed soybeans sat just outside the building. They shimmered bright green in the midsummer sun.

Inside the lobby there were more green things. Avocado-colored walls bore the company's slogan, "Innovation Through Nature."

Just past the lobby, Mac Orcutt, one of the company's meat specialists, greeted me. Orcutt was in his late fifties, with a broad forehead and narrow eyes that were often cast shyly toward the floor. He had just hauled down from his office an armful of empty boxes—each of them once containing foods made with soy protein—and was setting them up on a long table against the wall. There were boxes of Healthy Choices frozen dinners, Banquet frozen chicken patties, Morningstar veggie burgers, and plastic packs of Starkist tuna.

An Indiana native who once wanted to become a veterinarian, Orcutt went into animal science when he didn't get into vet school, later becoming a food scientist specializing in meat. Before Solae, he was at a company selling casings for hot dogs and pepperoni. His career continues to center on meat, though now he's focused on creating ways to mimic and supplant it. Sitting at a large conference table, Orcutt explained why soy protein is so effective for bulking up meat. Soy-infused chicken and beef, he said, often stay moist and juicy longer, which is a necessary consideration for industrially produced meats going through what Solae calls "high-abuse circumstances." Precooking, freezing, microwaving from frozen, and extended hold times in restaurants all serve to zap both moisture and flavor. Soy protein adds positive attributes without sacrificing any protein, as starches would.

Our lunch demonstrated the point. I was presented with a tasty and (thanks to the amazing properties of soy) impressively beef-like veggie burger, and two Cajun-seasoned chicken tenders—one of which had been simply marinated and baked, and another that had been extended 25 percent beyond its weight with soy protein and water. I was supposed to determine which chicken tender was which. When I tasted them, one seemed a bit more moist and easier to chew—the soy protein tender, I guessed correctly. Yet there wasn't a huge difference. Orcutt explained this was because both versions had been recently prepared in Solae's kitchen—cooked instead of industrially processed. Thus there'd been less opportunity for the meat to turn leathery.

Beyond taste and texture, the other big selling point for soy protein is its price. Soy protein is much cheaper than animal protein, and the water it absorbs is far cheaper than that. In a chart on its Web site, under the headline "Maximizing Profitability," Solae points out the relative price stability of soy protein compared to beef. Several years ago, the company helped a "large, specialized processor of red and white meats" to cut down on the amount of chicken it needed in its frozen cutlets. Using soy protein concentrate, the company was able to trim 11 percent of the meat for up to a 25 percent savings in raw material costs (versus non-enhanced meat), which counts as a lot of extra cash in the food business. And best of all, to customers, the new cutlets were indistinguishable from the old ones.

Using soy to amplify meat is a major factor in the national school lunch program. The school reform movement championed by outspoken chefs Jamie Oliver and Ann Cooper has promoted the idea of schools doing their own food prep and cooking from scratch. Some districts around the country are now actually doing this. Most are not, for reasons having a lot to do with cost. Under a system of free and reduced lunches that's been going on since the 1940s, the federal government currently reimburses schools between $2.46 and $2.86 per meal for each qualifying low-income student. Kids who pay for their own school lunch do so at about the same amount.

Such meager payments don't allow for the luxury of real working kitchens and the extra personnel that actual cooking requires. Nor do they account for the fact that fresh ingredients can go bad and sometimes have to be thrown away. Most school foodservice directors find it infinitely easier simply to reheat the inexpensive processed items that kids love and that food manufacturers are only too happy to sell. "We have to be fiscally responsible," is how Tony Jorstad, nutrition services supervisor for the Brighton, Colorado school district, put it when I visited one of his cafeterias. We spoke about Boulder County's program of scratch cooking, undertaken by Ann Cooper, who is dubbed the "Renegade Lunch Lady." Jorstad marveled: "In 2009, they lost $360,000. If I did that, I would lose my job." Boulder's school lunch program has since broken even, Cooper says.

Soy protein is one of the ways meat companies are able to sell bargain-basement-priced products to schools and still make a profit. Tyson, the country's largest meat processor and biggest provider of poultry products to schools, has been part of the USDA lunch program since 1984. Over the years, its school service division has developed quite a bit of proficiency in what it calls "chicken reprocessing." On its K-12 Web site, the company boasts that it does this better than anyone:

> When it comes to chicken reprocessing, Tyson is no rookie. Hands down, our process gets more servings out of your government allocation. Higher product yield results in more product for your money. With a variety of great-tasting, all-white and all-dark meat options to choose from, you'll be able to maximize every truck of diverted chicken.

That last "truck of diverted chicken" bit refers to the whole, raw meat that schools could get for free from the USDA if they wanted to. Instead, many schools opt to "divert" this free food to processors like Tyson so it can be turned into dinosaur nuggets, something they're actively encouraged to do by the USDA. Foodservice directors like Jorstad don't want the raw meat or whole potatoes because the cost to cook them and, in the case of meat, handle it carefully, would amount to more than buying the preprocessed stuff that's ready to pop in the oven.

Soy protein deserves some of the credit for these cost savings. Along with water, it replaces a portion of the pricier meat in Tyson products like Krisp N' Krunchy Chicken Patties, BBQ Chicken Chips, and Popcorn Chicken Bites. How much meat is being lost is hard to say. A USDA rule used to limit "vegetable protein product" in school lunch meat to no more than 30 percent, but after lobbying by Solae, it was removed in 2000.

The Journey toward Hotdogs

To understand how this versatile, cost-cutting, moisture-trapping meat stand-in is made, I drove five hours south, past more vast stretches of corn and soy to the outskirts of Memphis. Here Solae operates one of its five U.S. soy protein plants. Built in the seventies, the plant sits at the end of a tributary of railroad tracks that follow the Mississippi northward into Illinois. Each morning, railcars head south with loads of soybean flakes left over from the oil extraction process, some portion of the 2 or 3 percent that doesn't go for animal feed.

Ken Carnahan, a tall, athletic fifty-eight-year-old, is the plant manager, a position he'd been promoted to the week before my visit. A thirty-six-year veteran of Solae, Carnahan had a relaxed, confident air about him, as if he'd been plant manager for much more than a week. He wore a white hard hat, purple golf shirt, and form-fitted khakis to which he'd attached a large flashlight. His sturdy brown work boots were covered in a pale-yellow dusting of powder. "Soy flakes," he said with a shrug. "From when they come in off the railcars. You can't help it; it gets everywhere."

I put on a pair of mid-calf rubber boots, a set of earplugs, safety glasses, and a white hard hat much like the one Carnahan was wearing. Outside, at 9:00 AM, the temperature had already climbed to 90 degrees. Inside the plant, it was at least ten notches hotter, and I could see why. Three stories of steel boxes and vats hissed and whirred. Beyond the roar, it was hard to hear what Carnahan was saying, even after I took out the earplugs. The boxes and vats were connected by a serpentine system of pipes and tubes that steamed and sputtered. Inside that assembly of metal, soybeans with 35 to 40 percent protein were being turned into soy protein isolate with upwards of 90 percent protein.

To accomplish this, the beans need to be further subdivided. First, the fiber comes out. The flakes are mixed with water, which is then spun off in centrifuges. "You're looking at four million dollars' worth of machines right here," Carnahan shouted as we walked by a row of four. The discarded fiber goes to animal feed. Next, the water and soy flakes are sent through pipes into a giant acid tank, where they're mixed with hydrochloric acid and then centrifuged to get rid of the carbohydrates. Carnahan described this process as akin to adding vinegar to milk, which would curdle the milk and cause the solids to separate. But hydrochloric acid isn't exactly vinegar. It's found naturally in your stomach, but you'd never want to put it in there intentionally. It's highly corrosive and is used to remove rust or iron oxide from metals and to dissolve rocks in order to release petroleum and natural gas.

I understood why I was wearing rubber boots. The surface under our feet was slicked with water and dotted with occasional puddles of thick, foamy, yellow slurry dripping from a pipe. It smelled as if someone had taken animal manure and combined it with a pile of aging meat and old wet socks. If I closed my eyes, it would be easy to imagine a cattle feedlot, not a facility making the primary ingredi-ent in infant formula. Much of what's produced in Memphis goes to companies such as Mead Johnson and Abbott Laboratories for their soy-based baby formulas. I later asked Carnahan about the smell, and he said it was due to the release of sulfur compounds during processing. "After we spray-dry it, the smell is gone. Other-wise we wouldn't be able to sell a pound." He smiled.

After the acid bath, the soybeans are sent to have their pH lev-els neutralized with sodium hydroxide, which Carnahan likened to baking soda. But sodium hydroxide doesn't go into homemade Christmas cookies. It can cause blindness if it gets into your eyes, and its most common household use is as the active ingredient in Drano. In the food industry, it's also used for washing fruits and vegetables, for making soft drinks, and for removing chocolate's natural bitterness.

The sodium hydroxide is spun off in more of those pricey cen-trifuges, and the soybeans—or what were once soybeans—are piped off to be spray-dried. This is done by a row of machines powered by several 150-horsepower motors (each of which contains enough energy to propel a speedboat). They shoot the soybean sludge up toward the ceiling and through a tiny nozzle. The sludge flies out in tiny droplets into a large, heated bin, drying instantly into a fine powder. All of it happens in a flash. Dozens of gallons of soybean solution are "atomized" every minute, moving so fast that were anyone to stand in front of the spray, it would kill them.

Once dried, the powdered soy protein is transported via pipes to the packaging area, where it's portioned into 20-kilo bags, a steady stream of which zip around a conveyor to be sealed up and stamped with a sticker. As with soybean oil, the finished product is designed to be virtually tasteless, though this feat has proven to be an equal challenge. Taste issues have certainly improved since the early days—the sample Carnahan gave me to try had a very mild, slightly sweet, earthy taste. But when subjected to heat and moisture and mixed with other ingredients, those same undesirable tastes and flavors that once made soy protein inedible can reemerge. Raw soybeans have thirty-three different flavor compounds, and processing has a way of unleashing them. "Grassy, beany, and barnyard," said one sales manager at a company purporting to sell a "clean taste profile" soy protein. He added, "Someone once said it smelled like wet feet." Food manufacturers often address this by using what are known as "soy making agents," flavorings that neutralize or cover up those wet socks or whatever other funk might emerge.

A Nutrition Mystery

Soy protein isn't a harmful ingredient. Whether it's beneficial is another story. So, too, is the question of whether people actually want to be eating it, since many of us probably don't realize we are. If given the choice, most nonvegetarians (which is most of us—non-meat-eaters comprise no more than 3 percent of the population) would choose chicken and beef over a soy ingredient every time. There's a reason it's not called KFC's Country Fried Soy. Chicken and other meats are familiar food items that can be traced back to a farm without too much trouble. That farm may be a horribly smelly industrial operation with hundreds of thousands of chickens with inhumanely large breasts, but still—a farm.

You will have much more trouble doing this with isolated soy protein, since the journey will take you to at least two processing facilities—one like ADM's in Decatur and one like Solae's in Memphis—that have subjected soybeans to a series of disfiguring, chemically induced alterations. This is why the Clif Bar company, despite its best intentions, has never been able to produce a fully organic Clif or LUNA bar. The Bay Area–based company says it is committed to wholesome ingredients and environmental responsibility, but making a large, inexpensive supply of soy protein isolate, one of its main ingredients, according to chemical-free organic standards is tremendously difficult.

While in St. Louis, I asked Michele Fite, Solae's vice president of global strategy, whether people might feel unhappy about soy tucked into their chicken sandwich. She had a quick answer. "The people that are eating things like Banquet frozen meals aren't concerned about this," she said, motioning to Orcutt's boxes. Then, shifting to a different argument, one probably more in line with Solae's official messaging, she added, "Soy isn't a cheap filler. It's a tremendous protein alternative and it's something that provides nutrition. I'm incredibly passionate about this. The goodness of soy is so good for us."

Fite emphasizes that soy is a complete protein much like meat, containing all nine essential amino acids. Yet unlike meat, it has a relatively small environmental footprint. Soybean crops produce at least twice as much protein per acre as any other major plant species and fifteen times more than land used for meat production, all while consuming less water. Solae calls soy "the sustainable protein," ideal for feeding a growing global population.

But beyond protein, Solae's products don't have a lot to recommend them nutritionally, thanks to the prodigious amounts of processing they undergo. Whole soybeans are a tidy package of fiber, calcium, iron, potassium, folate (the natural form of folic acid), and vitamins B1, B2, B6, E, and K. They also have phytosterols and antioxidants such as saponins and phenolic acids. Whole soy foods like tempeh, edamame, soy flour, and tofu retain many of these nutrients. A 4-ounce serving of tempeh, for instance, has 50 percent of the daily recommended intake of fiber, 20 percent of iron, 11 percent potassium, and 15 percent calcium.

Yet by the time soybeans have gone through the manufacturing labyrinth to become soy protein isolate or soy protein concentrate (which is 70 percent protein) all of the fiber and vitamins have been processed out, either removed or destroyed by heat and chemical treatments. So have phytochemicals called isoflavones, compounds that may or may not have anticancer properties. The only things to survive besides protein are some of the iron, potassium, and manganese.

Commercial soy milk, which is often made from whole soybeans but goes through numerous processing steps, also doesn't have much nonprotein nutrition. A serving of soy milk delivers less than 10 percent of the recommended daily intake of iron, folate, zinc, and selenium, and as a result, most brands are fortified with what tends to be a telltale sign of shabby nutrition—a cocktail of synthetic vitamins and minerals.

There are few foods with a more conflicted and confusing nutritional scorecard than soy. According to the outdated yet persistent low-fat logic, soy protein is healthy because it contains little fat and no cholesterol. It's the primary ingredient in many products that sporty people like to eat, like Clif Bars and Odwalla Protein Smoothies. Many consider veggie burgers better for you than hamburgers, owing to their minimal fat content and perhaps the descriptor "veggie." Yet hamburgers look like Brussels sprouts compared to the nutritional vacancy of soy protein.

And then there's the curious question of heart health. In 1999, after a petition by Protein Technologies, now also part of Solae, the FDA gave the go-ahead for food makers that include enough soy protein in their products to put a heart-health claim on them. On foods like Kashi's GOLEAN cereals, you will see this message: "25 grams of soy protein a day, along with diets low in saturated fat and cholesterol, may reduce the risk of heart disease."

In considering Protein Technology's petition, the FDA focused on fourteen studies where researchers had fed people varying amounts of soy protein and then measured their blood cholesterol against control groups. While the agency concluded that this body of research did in fact show that consuming soy protein had a beneficial effect on cholesterol levels, there were several caveats. First was that soy protein only lowered LDL cholesterol (it had little effect on HDL, the good cholesterol) and it did this only when accompanied by diets low in saturated fat and cholesterol. Furthermore, it was only people with severely high cholesterol levels who saw significant reductions in their LDL. Despite all this, the FDA felt that soy protein still warranted a heart-health claim. The only real problem was that nobody could figure out why. "The evidence shows a clear relationship between soy protein and reduced risk of CHD [coronary heart disease] despite lack of a clearly defined mechanism for its effect," the agency wrote. This mechanism is still a mystery. Solae acknowledges it doesn't understand it, though it continues trying to figure it out.

This FDA-sanctioned link between heart disease and soy protein went more or less unquestioned until 2006, when several members of the American Heart Association (AHA) started to have doubts. Initially, the group had supported soy's heart-health claim and filed comments with the FDA saying so. But when a committee was assembled to reevaluate the science and look at some of the more recent studies, the results weren't encouraging. In a 180-degree reversal that highlights the imprecise and often flawed nature of nutrition research, Daniel Jones, then the AHA's president, wrote a letter to the FDA in February 2008, "strongly" recommending that the agency "revoke the soy protein and CHD health claim."

Jones argued: "There are no evident benefits of soy protein consumption on HDL cholesterol, triglycerides, lipoprotein, or blood pressure. Thus, the direct cardiovascular health benefit of soy protein or isoflavone supplements is minimal at best." His appeal failed to sway the government, which has yet to change its position on the claim, and Solae isn't backing down either. In a phone call, Ratna Mukherjea, Solae's global nutrition team leader, said that while she greatly respects the opinions of the AHA, the company has a different perspective. "What's important to me as a scientist is the consistency of the data that's been coming out since 1990, and we're still seeing the same information," she said.

Besides heart disease, the other unanswered health question that surrounds soy is cancer. Asian populations have had historically lower rates of hormone-related cancers such as breast, uterine, and prostate, and it's been suggested that their relatively high intake of soy might be a reason for this. Emboldened by the prospect of another gold star on its health resume, in 2004 Solae filed an FDA petition for an anticancer claim. It withdrew it less than two years later due to the fact that scientific research on the matter is deeply conflicted. And most soy protein doesn't retain a lot of the potentially anticancer isoflavones, anyway.

While some studies demonstrate a chemopreventive effect from soy (though perhaps only when consumed during adolescence), others show none at all, and still others suggest that isoflavones can feed tumor growth. Part of the confusion may originate from the fact that the term "soy" is so broad as to be virtually meaningless. Very little research has sought to differentiate between the bean's many iterations—the whole-bean fermented products like tempeh, edamame, natto, and soy sauce; the whole-bean nonfermented items like tofu, soy flour, soy grits, and soy milk; and the highly processed creations like soy protein, soybean oil, and isolated isoflavones sold as supplements.

Those who have sought to parse out soy's multiple personalities have arrived at suggestive results. In 2004, William Helferich, a nutrition professor at the University of Illinois and someone who has studied soy isoflavones for the better part of two decades, fed groups of mice with preexisting breast cancer several different soy diets. Each diet was carefully calibrated to have the same levels of isoflavones. His results showed that the least-processed soy foods, in this case whole soy flour, which is finely ground de-fatted soybeans, did not cause the mice's breast tumors to grow. Isolated, purified isoflavones, on the other hand, stimulated tumor growth. Helferich surmised that the disparity was likely to be a function of various bioactive substances in less-processed soy that cancel out the isoflavones' negative effect on cancer cells. "There's something to be said for complex foods for our bodies," he told me. "When the whole food is consumed you get a very different effect than if you consume the highly concentrated extracts."

Undoubtedly malnourished people, especially children, in developing countries would benefit from lots of soy protein. Regardless of where it comes from, protein is an essential nutrient for growth and something we all need to eat ample amounts of daily. (Most of us in the United States are already getting adequate amounts of this nutrient thanks to our heavy meat consumption.) But it's hard to see soy protein as an ideal answer to the world's hunger problems. What our bodies need along with protein is vitamins, minerals, phytochemicals, fiber, enzymes, healthy fats, and the whole complex array of nourishment found in foods that haven't been beaten to death. Exporting highly processed substances like soy protein to the developing world only slaps the problem with a Band-Aid. Eventually what you end up doing, if this plan of introducing "healthy" convenience foods from the First World to the Third is taken far enough, is swapping classic malnutrition for the modern diet-related maladies we have here in the United States. This arrangement benefits nobody except the food companies.

Liquid Chicken

One day, not long after I began research for this book, my husband arrived home from the supermarket with yet another product that piqued my curiosity. He couldn't find our usual brand of Bell & Evans breaded, frozen chicken tenders we buy for the kids, so he got a different variety. Like so many frozen concoctions, these tenders are superconvenient for working parents and eagerly wolfed down by the kids. I like the Bell & Evans variety in particular because they have a thick, meaty texture that makes them seem a bit closer to the real thing. Instead of these, my husband got Applegate Farms Organic Chicken Strips.

A few days later, after I heated up a serving, I noticed that the strips, which looked more like nuggets, seemed kind of puffy. I tasted them. The texture was totally different from what I was used to. It was airy and spongy, not very meaty. This chicken struck me as highly processed, yet when I looked at the box it informed me that the chicken I was eating was, in fact, "minimally processed." The listed ingredients seemed simple enough—organic chicken, water, organic rice starch, sea salt, and natural flavor in the chicken. In the breading, there was wheat flour and a bunch of flavorings.

Intrigued and confused, I did what by now had become an occupational habit. I added the remaining frozen nuggets to my collection of aging food items. Since they were organic, "minimally processed" and contained meat, I was prepared for an awful smell contaminating my office. Instead, I got something much more surprising. When I returned home from a trip about ten days later, I discovered that the Applegate nuggets, which I'd placed in a Ziploc bag left slightly open, no longer looked like chicken. Half of the contents of the bag had essentially liquefied, with the outlines of the individual chicken pieces no longer visible. The whole thing was soft and mushy to the touch, and the color had darkened. A few days later, the other half of the chicken had liquefied like the rest. The nuggets had completed their dissolution, and now all I had was a runny, brown mess.

Although in early 2012 pink-slime beef became a poster child for distrust of an industrial food-processing system, chicken actually endures considerably more high-tech poking and prodding than beef. Added soy protein is hardly the only thing we're talking about. Despite appearances, the chicken you find in the frozen aisles at the supermarket is almost never the same thing you would prepare at home. Those frozen nuggets, tenders, and breasts may sometimes start off as recognizable cuts of meat and use familiar ingredients, but then machines take over. More often than not, chicken is mixed under high pressure and tumbled together—under "high-abuse circumstances"—with a varying collection of other ingredients such as flavorings, starches, sodium phosphates, and soy protein. Then it's fashioned into tenders, nuggets, patties, boneless "wings," and "breasts." Even in cases where the meat is advertised as "whole muscle," sodium phosphates help the meat take on water, partly for profit yield and partly for antirubber insurance.

Chicken, it turns out, is never just chicken, something that's doubly true for any specimen you might find at a fast-food restaurant.

But this wasn't supposed to be the case with my Applegate strips. They were designed to look natural and wholesome, and it said so right there on the box—"all natural formed and breaded organic chicken" and "minimally processed." Eager to figure out what was going on, I called up Applegate, and several weeks later I had a perplexing conversation with Chris Ely, one of the company's founders. Before I'd gotten a chance to ask him about the liquid nuggets or their airy texture, he explained to me that Applegate was definitely "not in the sponge business." He wished to distinguish his product from so many of the other frozen nuggets on the market. "When you bite into our nuggets, you'll notice that our meat is a little loose in the center. We don't want to be a tight, bound-up product like a hot dog," he said. Other, more conventional manufacturers, he continued, mix their product excessively, using various additives to help add in more water and bind everything together snugly, lowering the cost.

When I got around to telling Ely about my experiment, he said that while he'd never tested for that sort of thing (who would?), his chicken might be more prone to disassembly because it isn't bound together with additives. Yet I later tested Tyson nuggets—which do have sodium phosphate and other additives—in the same manner and got the same result. (I also put the Bell & Evans tenders to the test and got foul-smelling chicken that remained intact. After two weeks, I couldn't stand the smell and had to throw them out.) I wanted to believe him. Applegate is an independent company and much of the meat they use is organic, with the rest coming from animals raised without hormones or antibiotics. The company is a big supplier to Whole Foods.

But animal muscle just doesn't liquefy unless something dramatic has been done to it. I told Ely that the combination of the brown mess in my office and my perception of the spongy texture seemed to suggest that his product was more maximally processed than minimally.

"I can see your point," he said, and then proceeded to launch into a largely unrelated discussion of mechanically separated meat, a product made by pushing a leftover, nearly meatless animal carcass through a giant tea bag–like screen at pressures so high that the whole thing—the skin, bones, connective tissue, and all—turns into a "meat" smoothie. Mechanically separated chicken (and pork and turkey; beef can't be "mechanically separated" due to concerns over mad cow disease) is the most industrial of meat concoctions, used commonly in hot dogs and pizza toppings. A widely distributed photo of it that showed a Barbie Doll–pink, serpentine substance oozing like soft-serve strawberry ice cream from an unidentified piece of machinery was initially mistaken for pink slime. Ely wanted me to know that Applegate doesn't use this stuff.

It's hard to say what exactly might have caused the company's meat to turn into goop. It may be related to the fact that its chicken strips go through an extruder to partially cook them and form them into identical pieces. As Ely explained it, the product my husband bought is made first by hand-deboning a whole chicken and coarse-grinding it in a mixer along with water, salt, rice starch, and oregano extract. Then the mixture is fed into the extruder ("It kind of reminds you of Play-Doh fun factories," he said). After that comes a breading device, a fryer, cooking in an oven, and finally freezing, all of which allow the nuggets to remain palatable for as long as a year, if the product is stored properly.

Despite Ely's best intentions, which I don't doubt, Applegate's nuggets/strips may simply be a telling illustration of the limits of modern food assembly. It just might not be feasible to make packaged frozen nuggets that resemble real chicken without dialing back some of the industrial manipulation, such as multiple cooking steps. Bell & Evans tenders are sold raw. They're subjected to one quick frying to set the breading, instead of Applegate's three heating steps. And there's no extruder; the chicken pieces in the box are from whole pieces of meat and come in different sizes and shapes.

Ely considers the sacrifice of some degree of authenticity to be a worthwhile trade-off in the name of certainty. "In today's food safety world, uncooked really scares me. Food safety is our number one priority, and we're not going to put out a product that could be accidentally not cooked properly," he said. It's a reasonable concern, though Bell & Evans has never had a recall of its products, and our family has never gotten sick from a chicken nugget that required thirty minutes of cooking in our toaster oven.

While the benefits of a safe food supply are undeniable and readily apparent, it's the trade-offs they sometimes come attached to that can be harder to grasp—especially when the word *organic* appears on the package. Those Applegate nuggets certainly weren't the worst thing in the world. But each nugget was more air and less chicken, and to me they felt more like a quick, forgettable fix than a satisfying meal. I figured maybe the kids would eat more of them to compensate, but that didn't appear to happen. If they had, though, they would have been filling up on a greater proportion of breading, which was not the point of the meal. Nobody answers the question of *what's for dinner?* with the answer *breading*.

Such compromises are quietly embedded into so much of our

processed food. They may come in the form of inadequate nutrition, the presence of less actual food than we think, or fleeting satiety, or maybe some degree of all of them. For most of us, some degree of these trade-offs is a necessary factor in the calculus of modern life; we give something up and get something in return. It's a useful swap—as long as we realize we're making it. The trouble with processed food is that it's rarely clear what exactly it is we're eating.

9

Why Chicken Needs
Chicken Flavor

*We are living in a world today where lemonade is made from
artificial flavors and furniture polish is made from real lemons.*

—Alfred E. Newman

There are few flavors that Marie Wright hasn't made in a lab,
or at least tried to. She's done sun-dried tomato, pâté, tacos,
lychee, and dog biscuit. She's created cheese flavors so realistic,
you'd swear you'd just eaten a square of blue or Monterey Jack.
Her biggest achievement, perhaps, was a white truffle flavor. She
calls it "completely authentic" and "very provocative." Wright grew
up in Essex, England, born to a Turkish dad and half-Italian mom.
Influenced by the feast of cultural cuisines she was exposed to at
home, she wandered into food science after studying chemistry at
King's College, London University. Her first job—working for the

ingredient giant Tate & Lyle doing sensory evaluations of its artificial sweetener sucralose—quickly became dull. She moved to New Jersey to work in the flavor business for legendary firm Bush Boake Allen and then for International Flavors and Fragrances, which acquired Bush Boake Allen. In fall 2011, she ascended to the job of chief flavorist at Wild, a rapidly expanding German company.

We met for lunch at one of her favorite restaurants in New York, just across the river from her offices in Elizabeth, New Jersey. She came bounding up the stairs with the energy of a teenager. In her early fifties, short and slender with long black hair and dark eyebrows, she was wearing floral capri pants and a gunmetal-gray cropped jacket. We'd come to a place she'd mentioned not long ago during a webinar she gave on global flavor trends. "If you ever have the opportunity to be in Belgium, Paris, or New York City, please go to Rouge Tomate," she said to her virtual audience of food and beverage product developers. "This is a fantastic restaurant and bar. Their list is inspirational and is also being used by large beverage companies to give them ideas for new products."

An elegant oasis in midtown Manhattan, Rouge Tomate is an offshoot of the popular and cutting-edge Belgian restaurant of the same name. Everything on the menu is phenomenally delicious, fresh, and incredibly healthy, in part because all the ingredients in executive chef Jeremy Bearman's kitchen are local and seasonal. Lettuce, tomatoes, herbs, and berries have often been plucked from the earth mere hours before they appear on your plate. The seafood is fresh from the oceans. The chicken on your plate was ambling around a farm not more than three days ago.

Behind the bar, you won't find cranberry juice or bottled ginger beer—they make their own. Everything gets hand-muddled, squeezed, and carefully blended into a seasonally evolving menu of drinkable concoctions, many with unusual pairings of flavors that

Wright finds enthralling. She sometimes draws inspiration from the restaurant's drink list for products she develops for beverage clients.

When the meal arrived, we admired it for a minute before eating. "This is just beautiful," she said with a British accent, sliding a thin, barely cooked slice of Arctic char into her mouth. "Food that's fresh has an incredible taste. A kind of cleanness."

We ordered a few drinks to try. One was called Red Roots, a frothy, intensely purple mix of beets, carrots, cucumbers, lemons, and ginger. "Wow, that's a lot of beet," Wright said, after sucking some up through a straw. "It has a strong earthy note, like the taste of dirt." She added: "There's a big interest in dirt right now, mostly for high-end baked goods. General Mills isn't asking for dirt flavor in their brownies yet, but these things can trickle down." She noted with delight Jack in the Box's experimental bacon milk shake and Burger King's bacon sundae. Wright also sensed a lot of "fatty" notes coming from the drink's cucumber, a sensation I never would have arrived at on my own. Flavorists are able to pull apart what to most people seems like one or two tastes into sometimes a dozen different "notes."

It's not uncommon for Wright to find flavor inspiration from celebrated chefs. Not long ago, she saw a recipe for passion fruit and basil macaroons from Adriano Zumbo, an Australian pastry chef. It occurred to her that basil's anise and floral notes might add a certain twist to the taste of passion fruit. She envisioned making not a basil–passion fruit flavor, but something with a slight twist on the latter's sweetness, potentially a perfect flavoring for fancy chocolates. "All good ideas are stolen," she observed wryly.

Despite the ingenuity they offer, Adriano Zumbo's precious $10 macaroons and Jeremy Bearman's seasonal offerings at Rouge Tomate are light-years away from the products most of Wild's flavorings end up in. Manufactured tastes are one of the most defining ingredients in processed food, as prevalent as sugar, salt, fat, wheat, and synthetic vitamins and minerals. You almost can't find packaged food in the supermarket or fast-food ingredient lists without the words "natural flavors" or "artificial flavors." These taste materials make the entire beverage industry possible, as well as most ice creams and yogurts. They lend cravability to foods that would otherwise be inedible slurries of corn, wheat, or soy—products like Doritos, some breakfast cereals, and energy bars. And they allow manufacturers to put less fruit into cereal bars and pie fillings, with strawberry and apple flavoring making up the difference; to make juices from raw materials that aren't very fresh; and to deliver the impression that processed meats have been cooked the way you would cook them at home. You can see this with supermarket brands of frozen pre-"grilled" chicken breasts. In lieu of actual grilling, flavor companies like Wild give them a nice, savory, charred "grill" flavor that was most likely made by processing vegetable oils at extremely high temperatures. To complete the impression of grilled chicken, a machine imprints them with blackened strips and caramel coloring gives the outer surface a golden glaze.

I wondered how Wright, someone with a heightened reverence for simple, wholesome, garden-fresh foods, felt about participating in the manufacture of fake grilled chicken breasts and fruit bars without any fruit. "If I can make a bland piece of chicken taste better, then I'm fine with that," she said. "Because that chicken is still good protein and nutritious. Not everybody can afford or have access to real, fresh foods."

I asked Wright about foods that aren't chicken, all those nutrition-free products that flavors help make us crave—sugary drinks, snack chips, puddings. "Are we responsible for people's will power?" she asked rhetorically. "We're just making it taste good."

Wright is a devoted mom of two school-aged kids, and she is passionate about her career. "Few things make me happy like creating flavors," she said. "I can't imagine not doing it." A few years ago she oversaw an edition of the artsy publication *Visionaire* that featured celebrities like Yoko Ono and surfer Laird Hamilton conceiving of concepts that flavorists created scents strips for. Her favorites were "orgasm," an aroma of chocolate, truffles, and sweaty bodies, and "exotic," a mix of mango, orange blossom, and pepper.

The dissonance between her own diet and the processed, packaged creations she helps formulate for others seemed not so much an expression of cynicism or elitism but rather a symptom of the intellectual compartmentalization of the contemporary workplace. Wright derives meaning and livelihood from the art and science of flavor creation, and it's not necessarily within her purview to think about where those aromatic melodies end up, who eats them, and in what quantities. As she put it, her job is to make things taste good, something she's exceptionally good at.

Unhappy Chickens

Of the roughly five thousand additives allowed into food, over half are flavorings. These thousands of taste molecules serve not only as window dressing designed to make food hyperappealing, but often as the very foundation of the house itself. Consider KFC's gravy, a product with at least seven flavoring ingredients, or nearly a third of the total:

Food Starch-Modified, Maltodextrin, Enriched Wheat Flour (Niacin, Reduced Iron, Thiamine Mononitrate, Riboflavin, Folic Acid), Chicken Fat, Wheat Flour, Salt, Partially Hydrogenated Soybean Oil, Monosodium Glutamate, Dextrose, Palm and Canola Oils, Mono- and Diglycerides, Hydrolyzed Soy Protein, Natural and Artificial Flavor (with Hydrolyzed Corn Protein, Milk), Caramel Color (Treated with Sulfiting Agents), Onion Powder, Disodium Inosinate, Disodium Guanylate, Spice, Spice Extractives, with Not More Than 2% Silicon Dioxide Added as an Anticaking Agent.

This is an unusual example in the sense that you can identify most of the flavorings. More often than not, you can't. They are tucked into the opaque designations *natural flavors* and *artificial flavors*, which include things you can taste—fruits, spicy notes, savory, salty, and tangy flavors like lemon or vinegar—and substances you can't, because they're being used to cover up unwanted flavor. Many ingredients that go into processed food don't actually taste very good and need to be masked. In addition to soy protein, there's the bitter taste of most artificial sweeteners and preservatives like sodium benzoate and potassium sorbate, which impart what's known as "preservation burn." Wild has a product to modify the taste of stevia. "It has this horrible liquorice flavor that lingers," Wright noted. Added vitamins taste, unsurprisingly, vitamin-y. B1, in particular, can have a rotten-egg aroma.

A chef would make a gravy using poultry fat and stock, along with butter, onions, flour, cream, salt, pepper, and maybe white wine, but industrial processors, for the most part, don't have this luxury. Using real ingredients is not only more expensive, it's often ineffective, since Mother Nature's volatile and fragile flavors often don't fare well during journeys through the assembly line. The potions produced by Wild, International Flavors and Fragrances (IFF), Gividuan, the world's largest flavor company, the Swiss company Firmenich, the German outfit Symrise, Sensient, which is based in Cincinnati, and a handful of others are much more sturdy.

"If you take a fresh strawberry after processing, it's nothing. It tastes like nothing," said Wright, as a way of explaining why the food industry is so reliant on the $12 billion global flavoring industry.

Some of the demand for flavoring is related to how plants and animals are grown and raised. Wright urged me to try a taste test at home if I was so inclined. Take three different whole chickens, she said—an average, low-priced frozen one from the supermarket; a mass-produced organic version like Bell & Evans; and what she termed a "happy chicken." This was a bird that had spent its life outside, running around and eating an evolutionary diet of grass, seeds, bugs, and worms. Roast them in your kitchen and note the taste. The cheap chicken, she said, will have minimal flavor, thanks to its short life span, lack of sunlight, and monotonous diet of corn and soy. The Bell and Evans will have a few "roast notes and fatty notes," and the happy chicken will be "incomparable," with a deep, succulent, nutty taste. Wright, as you might imagine, prefers consuming chickens of the happy variety, which her husband, who is also a flavorist (he works from home as a consultant) is generally the one to cook.

I already knew a little bit about this. Several months before meeting Wright, I'd gotten a tutorial on how some of the flavors of "happy chicken" can be reproduced. In the middle of suburban New Jersey, I visited a company called Savoury Systems International, a small, specialized player in the flavor universe. As its name suggests, the company creates savory, meatlike flavorings for the food industry. Its offices are located in the front corner of a generic office park in Branchburg, and like all flavor operations with labs or finished ingredients on-site, the office was permeated with a smell. It evolved throughout the day. As I entered the building, it was sweet, fruity, and meaty, like someone was baking chicken nugget-flavored Lifesavers. Later, on my way out, it smelled more like hot dog buns. The Air Wick plugged into an outlet in the small lobby had been outmatched.

Inside the lab, just past the lobby, Kevin McDermott, Savoury Systems's young, eager technical sales manager, offered me a taste test. The first was a powder made from actual chicken parts. He mixed it with some warm water and poured it into two small plastic cups for us to sip. I contemplated the pale yellow liquid warily and then tried a bit. It was weak and washed out—a bit revolting. McDermott got out another plastic bin of powder, scooped out a little and mixed it in a beaker with warm water. This was hydrolyzed vegetable protein, or HVP. Made from soybeans, HVP is one of the core ingredients used to construct the company's meat flavorings such as "roast chicken type flavor" and "natural juicy beef flavor." The HVP liquid tasted great, like juicy chicken infused with soy sauce. It tasted a whole lot more like chicken than the real article. McDermott gave me another HVP to try, this one with a darker, almost burnt flavor. Also delicious.

Vegetarian substances like HVP and yeast extracts, which Savoury Systems also uses, can be made to taste exactly like chicken or beef because they mimic the flavor of meat on a molecular level. On its own, soy protein doesn't have any meat flavor, but cleave it apart into its component amino acids (which are the building blocks of all proteins), and dynamic flavors emerge. Some of them, like that from leucine and valine, are truly nasty; other amino acids trigger our taste buds in pleasurable ways. Glutamate, for instance, is the reason monosodium glutamate (MSG), is such a useful flavoring ingredient. Glutamate also imparts flavor to HVP and yeast extracts, though it's present at lower concentrations than in MSG.

When I asked Dave Adams, the food scientist who founded Savoury Systems, why actual meat is such an inferior source for the chicken flavor that, strangely enough, goes into chicken, he gave me the same answer Wright did. Modern chicken, he grumbled, has no flavor. "They grow them so fast, they don't have time to develop flavor," he said. And chicken—even tasteless, scrap stuff— is more expensive than soy.

To get HVP, a de-oiled soybean meal (or cornmeal) is boiled in large vats of hydrochloric acid for six hours, wrenching apart protein molecules into amino acids. Corn syrup can be added to the mixture to yield a more intense browning flavor. The solution is then neutralized with sodium hydroxide, which leaves the final product with an abundance of sodium. (In response to the food industry's emphasis on sodium reduction, some HVP makers have tried to produce lower-sodium versions, with varying degrees of success.) Yeast extracts are made in a similar fashion, although no chemicals are needed. Yeast cells are killed with an excess of salt and heat, triggering the organism's own enzymes to break down its protein into amino acids. Hence the term autolyzed (self-digested) yeast extract.

At IFT 12 in Las Vegas, I stopped by the booth of a flavor company called Innova to experience the magic of these flavors in actual food. Scientists, all wearing long, blue, collared shirts that matched the hue of the thick cerulean rugs underfoot, were serving up crock pots of what tasted like beef barbacoa. A dish that hails from the Caribbean, beef barbacoa is traditionally prepared by covering meat or sometimes an entire animal with leaves and cooking it in a hole in the ground until the meat is succulent and tender. In more contemporary kitchens, it's beef cooked slowly with broth and lots of spices. Chipotle, which uses a version of the recipe in its restaurants, describes it as "spicy, shredded beef, slowly braised for hours in a blend of chipotle pepper adobo, cumin, cloves, garlic, and oregano until tender and moist." Innova's barbacoa actually tasted a bit like Chipotle's, although less spicy. It was moist and savory and a bit sweet. I immediately went back for seconds. Only after finishing did I realize that what I was devouring wasn't beef.

The meat hadn't been slow-cooked for a day; it was cooked quickly, then frozen in a bag and eventually reheated. And it was chicken dressed up with a manufactured "barbacoa spice type flavor" and Innova's "natural beef type flavoring," consisting of hydrolyzed yeast extract and MSG. The forgery was acknowledged and intentional. It was designed to showcase Innova's abilities to help large food processors—who don't have time to slow-roast and whose factories would be unkind to spices like cumin, cloves, and oregano—take money-saving shortcuts to get great-tasting meats, whether at restaurants or for meals in the frozen aisle. Innova's fake beef barbacoa tasted not exactly like the real thing, but close enough.

Mapping Nature

The flavoring game wasn't always so sophisticated. When it began in Europe in the nineteenth century, companies imported spices and procured plants such as lemongrass, which yielded citronella oil, ideal for concentrating into lemon flavor. These essential oils went mostly into fragrances, medicines, and candies. As the field of chemistry progressed in the latter half of the century, European scientists, particularly Germans, figured out how to synthesize flavors and fragrances from chemicals instead of having to wrench them from natural materials.

When the first flavor companies appeared on U.S. shores, they clustered along the East River in Lower Manhattan, near what used to be the Fulton Street Fish Market, within spitting distance of ships arriving from Europe with essential oils and synthetic chemicals. The area was so thick with scents that it became known as the "Aromatic Circle."

As it did in so many other areas of commerce, World War II forced transformative market changes when supplies from Europe and elsewhere were cut off. Many companies expanded and moved across the Hudson to set up new factories. New Jersey is still a hub for the flavor industry today. Gividuan manufactures products there, as does IFF. Wild is in Elizabeth (though its U.S. headquarters are outside Cincinnati) and Symrise maintains three New Jersey sites, one of them just down the road in Branchburg from Savoury Systems.

Postwar foods were flying off assembly lines, and they needed flavoring, so companies churned out all kinds of novel formulations. Dow Chemical created allyl cyclohexanepropionate, which it touted in ads as "fresh pineapple flavor." The Swiss company Firmenich developed the first strawberry flavoring and created a compound called Furaneol that would become essential to products like Jell-O and Kool-Aid fruit punch. The company described it as "a sweet, cotton-candy like molecule key to red fruit, tropical fruit, and roasted flavors." These and other compounds were supposed to give processed foods and drinks the same flavors as foods prepared at home, but they often fell flat. Flavors then were still vague approximations of the real thing. In 1952, *Fortune* magazine declared, "It is hardly surprising that, in the opinion of many, the flavor of American food and drink—in jars, cartons, bags, cans, fifths and pints—leaves something to be desired."

In nature, flavor comes as a sophisticated mix of hundreds, sometimes thousands, of chemicals, each with its own unique taste and smell. Using early-twentieth-century chemistry tools, scientists could hope to identify perhaps a handful of these in any given plant, and doing so was cumbersome and imprecise. Gas chromatography changed all that. These machines were first developed in the fifties and put into wide use by the seventies. They move gasses along a tube and isolate molecular constituents based on different boiling points and variations in their polarity. Coupled with mass spectrometers, which identify what's been isolated, this technology opened up a vast world of possibilities, allowing for a much more thorough (though still incomplete) map of nature's aromas. The number of flavor chemicals known in orange peels, for instance, has gone from 9 in 1948 to 207 today. In spearmint leaves, it's leapt from 6 to 100, and in black peppercorns, from 7 to 273.

Flavorists today can come close to capturing and approximating that elusive, clean taste of freshness that Marie Wright and I savored during our Rouge Tomate lunch. To do this, they trek to farms when a crop is at the peak of ripeness, taking portable gas chromatography devices with them. They drape the strawberry or tomato or pepper plants with plastic bags or glass jars, corralling the aroma gasses in an attempt to make an imprint. The aim is not to preserve the gasses, though; they're way too fickle. Back in the lab, you work on mimicking what the machinery has identified. Wild's scientists have done this sort of "headspace analysis," as it's called, in mint fields that the company operates in southern Washington state. And like most of its competitors, Wild sells "fresh" versions of many of its flavors, some more convincing than others.

One of the newer breakthroughs to come along in the science of flavor is called taste modulation. About a decade ago, a biologist at the University of California at San Diego named Charles Zuker isolated, for the first time, the receptors on the tongue that are responsible for our perception of taste. He did this using taste-bud cells from laboratory mice. What he found was that each cell was incredibly specific, containing receptors for just one taste—either sweet, sour, salty, bitter, or savory (also called *umami*). This was great news for the food industry. It meant that these taste-bud cells and their receptors could be much more easily manipulated than if they were being bombarded by multiple flavors. Zuker, who went to college at age fifteen and had his PhD at twenty-six, realized he had the tools to start changing the biology of taste. He founded a company called Senomyx.

Now a public company that has done deals at one point or another with Pepsi, Coca-Cola, Nestlé, Kraft, and Campbell's Soup, Senomyx's flavoring products and potential products are designed to block certain sensations like bitterness—a more targeted form of taste masking—or to heighten them, allowing companies to trim back their use of sugar, sucralose, salt, and MSG in products, while still preserving the sweet or salty taste. Senomyx says that items containing its sweet enhancers and savory enhancers, both of which have no taste themselves, are currently being sold in the United States, appearing on the label under the catchall "artificial flavors."

Not surprisingly, Senomyx no longer has the business of taste modulation to itself. All of the major flavor companies, including Wild, have similar research programs underway. Many of these enhancements are targeting the creation of healthier packaged foods, with less sugar, salt, and MSG. In an interview with *Scientific American* in 2008, Zuker, who is now at Columbia University and not involved with the day-to-day running of Senomyx, spoke about the formation of the company. "We thought, here perhaps we have an opportunity to help make a difference."

As Wright would put it, we all can't live at the top of the food chain, a place where meals aren't loaded up with excessive salt, sugar, and MSG. During our lunch, I got the sense that in a parallel universe—one where processed foods don't pay all the bills—she'd be designing potions mostly for experimental, fancy foods—those passion fruit chocolates and dirt-infused brownies. That is to say, foods she consumes and wholeheartedly cherishes. When I posed this idea to her over e-mail, she said that she would, in fact, one day love to have her own flavor studio for "the creation of exquisite tastes."

"That remains my goal when I do not need to earn much money."

10

Healthy Processed Foods

The modern donut . . . is vastly different from the old concept of the indigestible grease-soaked donut. . . . The baker exercises precision control on fat absorption, fat temperature and time in which the donuts are being cooked. We do believe that the modern donut is a highly nutritious, energy food.

—J. I. Sugerman of the Doughnut Corporation of America, in a 1942 letter to Wilburn L. Wilson, Assistant Director in Charge of Nutrition, Department of Defense, appealing for the company's "Vitamin Donuts"

It's 10:15 on a Tuesday morning and I'm crunching on a biscotti. I often find myself eating these hardened Italian biscuits with coffee as a quick snack. The beauty of them is that they deliver the promise of something sweet without the sugar assault of, say, a giant cinnamon roll or chocolate chip cookie.

But this biscotti was no ordinary baked treat. It was created as a prototype by a company called Penford and contained a special type of starch designed to function as dietary fiber. I sat eating it in the company's unremarkable conference room in Centennial, Colorado, and it tasted to me exactly like a biscotti I'd get at a coffee shop. It had a firm, crumbly texture and small slivers of almonds inside. There wasn't any trace of fibrous material, no gritty feel or branlike flavor. It seemed the 13 grams of fiber it contained were totally disguised, as if they weren't even there, which of course was the point.

Penford is one of roughly a dozen companies that sell starches to the food industry. Located just south of Denver in an office park so nondescript it could have been plucked from pretty much anywhere in the United States, the one-hundred-year old firm is a recent transplant to Colorado. Its roots are in Cedar Rapids, Iowa, where it was a leader in corn syrup production. A 1945 photo in a book on the company's history shows a "Penford Corn Syrup" truck pumping its contents through a hose into the side of a building, a bulk storage "filling station," in Cambridge, Massachusetts.

Starches arrived on the company's radar after scientists, working mostly for another company, National Starch, figured out in the fifties and sixties how to use chemicals like sodium trimetaphosphate and phosphorus oxychloride to modify starches in interesting ways. Penford sold these specialized starches to the paper and adhesive industries for several decades, and then discovered a lucrative market in the fast-food industry. In the eighties, a Penford scientist devised a way to use potato starch to help keep french fries from sticking together when frozen and to increase their crispiness after being fried. Today, several large fast food chains still dip their fries in a solution consisting of Penford's potato starch.

Besides forming a clear sealant on fries, Penford's products also help thicken sauces, make ice cream creamier, allow meat to retain water, extend the shelf life of tortillas, and do to baked goods what many conditioners claim to do for hair—add volume and structure. The company's starches originate from potatoes, tapioca, rice, and corn, and its corporate motto is designed to convey the idea that Penford is improving upon natural resources; its ingredients, the company says, are "Created by nature, advanced through science."

The fiber product in my biscotti, called PenFibe, represents the company's entry into the growing market for healthy processed foods. Instead of getting sucked up rapidly into the bloodstream, as would normally be the case with most starch, this so-called "resistant starch" has been molecularly rearranged to withstand human digestion—to be resistant to it. It can be easily and stealthily added to a variety of different foods, especially baked goods, to increase the number of fiber grams appearing on the nutrition panel. As both starch and fiber simultaneously, it suggests a whole new, ultraconvenient, compromise-free way to get more of this important stuff into our diets.

After all, fiber may be even more valuable than we realize. Along with its most infamous task of keeping us regular, it serves to slow down digestion and prevent the rapid release of glucose into the bloodstream, which helps maintain consistent energy levels and prevents conditions that can lead to type II diabetes. Fiber also plays an important role in lowering blood cholesterol and blood pressure and in reducing inflammation. Some types function as fuel for the healthy microorganisms in our gut, especially those in the colon, a dynamic that scientists believe can aid in the prevention of diseases like diverticulitis. If all that weren't enough, it's also thought that fiber can bind to cancer-causing chemicals in the colon, blocking them from inflicting damage to cells.

So we need fiber. But, by and large, we're not getting it. Fiber is one of those fragile constituents of food, like vitamins, phytochemicals, and flavors, that's easily obliterated in processing. Often, more than half of what's naturally present in whole grains, for instance, won't survive the heat, mechanical mixing, shear, hydrolyzation, and whatever else is being done to it before it gets sealed in a box or bag. This is one reason that many cereals, even those touting "whole grains" on the package, have some type of added fiber in their ingredients. The government recommends that adults get anywhere from 25 to 38 grams of fiber daily. Yet studies show that the vast majority of us are consuming no more than 15 grams per day (or half the amount our ancestors ate a hundred years ago).

Penford wants to help end this fiber famine. When I visited the company's offices in Centennial on a bitingly cold December day, Bryan Scherer, the company's head of technology, showed me around. He took me to the lab where company scientists experiment with new starches. Bags and tubs filled with white powders sat on big steel tables. In other rooms, scientists worked on prototype products. One room contained a pizza oven and proofing boxes for bread. Another had a machine for making sausages. Two workers wearing hairnets operated a machine that repeatedly stabbed whole chickens with eighteen needles, injecting them with a solution of salt, sodium phosphates, water, and starch. There was also a machine for making soft-serve ice cream. "Rice starch does very well if you want to replace the fat in ice cream. It can give you a product that's 95 percent fat free," Scherer noted.

Ibrahim Abbas, a senior research and development manager who studied food science at the University of Baghdad, and Jennifer Stephens, Penford's director of marketing, joined us in the conference room to discuss fiber. Scherer explained PenFibe's unlikely origins: potato chip and french fry factories. Penford collects wastewater from these plants and isolates starch that's been leached off potatoes. (The company touts this as part of its sustainability initiative; most starch sellers grow their supply.) The starch is then treated with chemicals to strengthen the joints between molecules so they can't be broken during digestion. (Resistant starches can also be made through a heating and cooling process.) After that, it's dried into a white powder that can last for upwards of two years.

Scherer ran though all the benefits of PenFibe: how it can replace up to 20 percent of the flour in baked goods, increasing fiber while reducing calories, since, pound for pound, it contains 80 percent fewer calories than flour; how it doesn't succumb to processing mutilation like the naturally occurring fibers in whole grains do; how it can't be detected by taste buds, which my biscotti experience clearly illustrated.

To get a sense of how much 13 grams of fiber is, here are other foods I'd have to eat to get the same amount:

> One pear, a large carrot, a banana, and half a cup
>   of broccoli
> One cup of black beans and brown rice

Before arriving at Penford, I'd stopped at Starbucks for oatmeal. That gave me only 4 grams of fiber. Now, after the biscotti, I had a huge jump start on my daily fiber intake. It sounded too good to be true.

Is It Fiber?

When resistant starches first started appearing in the food supply in the nineties, government authorities didn't know what to make of them. These substances registered as fiber when subjected to the official AOAC test for fiber analysis, a process that seeks to mimic human digestion in a lab. (Founded by Harvey Wiley and originally known as the Association of Official Agricultural Chemists, the AOAC is a scientific group that oversees various testing methods.) But they were also quite different from the produce, grains, and beans we normally associate with fiber. Who ever thought fiber would come from a Ruffles factory?

There were other types of new fibers, too: inulin, which hails from the roots of chicory plants; polydextrose, a synthetic derivative of glucose; a soy fiber from Solae, cast off from soy protein production; and soluble corn fiber, which is in Splenda along with all those vitamins. The FDA looked at the possibility of revising food labels to distinguish between naturally existing and manufactured varieties, an idea ingredient companies hated. They were worried the distinction might leave consumers with the impression that nature's fibers were somehow superior. The FDA backed off and, as it often has in the face of industry dissent, left the issue to the industry. The American Association of Cereal Chemists convened a ten-person panel consisting of four academic food scientists, four individuals from the food industry, a current USDA official, and a former FDA official to examine the matter. In 2000, the panel concluded that there should be no distinction between types of fiber; all new forms of it deserved equal voice on packages. This decision ultimately received the FDA's tacit approval; the agency didn't object but never published a formal rule. The definition was also adopted by the Australian and New Zealand governments and the European Food Standards Authority, Europe's FDA.

Not everyone was so sure, though. A panel convened by the Institute of Medicine in 2001 drew a clear—and controversial—distinction between naturally occurring fiber and isolated ingredients.

A group of experts assembled by the United Nation's (UN's) World Health Organization and its Food and Agriculture Organization went even further to conclude that manufactured fibers shouldn't be considered fiber at all, though several years later, after much worldwide objection, the UN group came around to the broader definition. Part of the initial reasoning of the UN experts, though, was a belief that human digestion can't actually be replicated in a lab. Supporters of manufactured fibers concede that while this is true in theory, we know enough about how digestion works, especially in terms of acids and enzymes, to copy the basics.

Yet the process of digestion is inseparable from the immense, still unfathomable collection of microorganisms in our guts. A number of companies selling isolated or manufactured fiber say they've done human feeding studies showing that their products are, in fact, beneficial to gut microflora, helping to promote their growth in the colon. Still, there was this note of skepticism in the USDA's 2010 Dietary Guidelines: "Fiber is sometimes added to foods and it is unclear if added fiber provides the same health benefits as naturally occurring sources."

The most obvious way that substances such as PenFibe are different from what's in brown rice and blueberries is the absence of other nutrients packaged with them. Nutritionally, I would have been much better off eating beans and rice instead of the biscotti. I also would have felt like I'd eaten a meal, not a passing sugary snack.

Bryan Scherer and Jennifer Stephens don't dispute this. Of course it's better, they acknowledge, to eat a banana than a fiber-enhanced, banana-flavored muffin, or to make a homemade shrimp bisque, as Scherer did the other weekend, than to eat fast food—but are people really going to start making food choices based on what's best for them? In Scherer and Stephens' estimation, the answer is: probably not. During my visit, Stephens told a story about a friend who works at a Whole Foods store. One day a man came in and said he needed help eating better because his doctor had warned him he was headed down a long, dark road. Thrilled, her friend led the man over to the store's bulk area, presumably to show him where the walnuts, granola, and quinoa resided. A worried look descended on the man's face, and he confessed he didn't even own a pan. All he could do was heat things up in a microwave and would she please just take him over to the freezer aisle and show him some microwave dinners that might be a bit healthier than what he usually buys.

Many Americans, in other words, perhaps the vast majority of us, are going to continue eating a steady supply of frozen dinners, buckets of fried chicken, muffins, bagels, donuts, and Hot Pockets, even if we know better. So the industry may as well make these craveable items a little bit healthier. "You really have to look at the consumer base," she noted. "People have busy lifestyles and they're eating prepackaged and fast food, and there's no way to get much [naturally occurring] fiber that way. These new types of fibers are a way for the food industry to help people out."

Much as Marie Wright doesn't consume most of the products her company's flavors show up in, Stephens, Scherer, and Abbas made it clear that they themselves are not "the consumer base." They manage to buy a good deal of fresh food and derive enjoyment from cooking their own meals. Abbas said he makes his own yogurt, and Scherer is as much a cook as he is a food scientist. Years ago, he took cooking classes at Johnson & Wales in Rhode Island and once started a salsa company based on his homemade recipes. He talks about the recipes he's constantly ripping out of magazines and reworking in his kitchen with as much authority and enthusiasm as he does cross-linked starches.

Everyone at Penford seemed genuine in their desire to offer helpful solutions, and I'm sure the fact that their company is going to make money doing so only adds to their conviction. They feel, as do many in the food industry, that maximizing sales and profits and stemming America's descent into chronic disease are not mutually exclusive propositions. After all, food companies have multitudes of brilliant people working for them and millions of marketing and research and development dollars at their disposal. Why not marshal some of this for the benefit of national health, particularly the physical well-being of our kids?

It's a heartwarming idea, and there are plenty of people who subscribe to it, including some very high-profile ones.

Unfortunately, it's not going to work.

"Better for You"

On a bright morning in March of 2010, Michelle Obama stepped up to a podium wearing what *InStyle* identified as a lacquered, high-waist pencil skirt and a bicep-showcasing pink L'Wren Scott halter top. Sprawled out below her, packed into the ballroom at the Grand Hyatt Hotel in Washington, DC, were nervous members of the packaged-food industry. This was one of Obama's first public talks on what would become her signature First Lady issue—ending childhood obesity within a generation—and no one was quite sure what she was going to say. Would she advocate for dreaded limits on kids' marketing? Would she talk about how fresh foods are inherently more nutritious than highly processed ones? Worst of all, maybe she'd call for a diversion of farm subsidies from corn, soybeans, and wheat toward the fruit and vegetable crops she was certainly going to be urging kids to eat more of? After twenty-five minutes, the answer was clear. Obama wasn't going to go there.

She would talk about how the food industry needed to change its ways, but in an accessible, friendly way. No one was going to get left out in the cold.

"We need you not just to tweak around the edges, but to entirely rethink the products that you're offering," she suggested. "That starts with revamping or ramping up your efforts to reformulate your products, particularly those aimed at kids, so that they have less fat, salt, and sugar, and more of the nutrients that our kids need." Such healthier packaged foods, Obama said, have enormous potential to "help shape the health habits of an entire generation."

She described how she personally understood the siren call of processed foods. "It wasn't long ago that I was a working mom dashing from meetings and phone calls, ballet and soccer and whatever else. . . . I bought products that were prepackaged, precut, precooked. If it wasn't 'pre,' I wasn't getting it," she said, to a round of relieved laughter. "Because I was looking for anything that was quick and easy to prepare and to consume. And I was grateful for the time and the effort that I saved with these kinds of products."

The food industry loved what they heard. When Obama was done speaking, the crowd stood up and flooded the room with applause. What the White House's new obesity and health czar had just said was that the packaged-food industry not only could, but *needed*, to be part of the solution. Companies had to cut salt, sugar, and fat from their products and add more nutrients like vitamins and minerals and maybe dietary fiber. This was criticism they could work with. In fact, it was the sort of thing they were already up and running with.

On its Web site, the Grocery Manufacturers Association (GMA), a group that represents companies like Kraft, General Mills, Pepsi, and Kellogg's, has an account of its successes. "Since 2002, GMA member companies have introduced more than 20,000 products with fewer calories and reduced fat, sugar and sodium," the group said. "And through the Healthy Weight Commitment, food and beverage companies have pledged to remove 1.5 trillion calories from the food supply by 2015."

It sounds impressive, but it's not going to matter. Despite Michelle Obama's ample accomplishments in getting everyone fired up over the issue of childhood obesity, her prescriptions for the packaged-food industry won't change America's dreadful eating habits.

One of the reasons for this is the loose way the industry tends to define the word *healthy*. Consider, for instance, a report done in 2011 by a former food-industry executive who now works for the Hudson Institute, a conservative think tank. The report detailed the way the sales growth of various "better for you" products has outpaced that of "less good for you" products over the past five years. (The food industry avoids the terms "unhealthy" or "bad" like the plague. Soda and Doritos aren't junk food; they're "fun for you" choices, a term PepsiCo coined.) According to the report, between 2007 and 2011, among the eight thousand packaged products evaluated, healthier choices made up roughly 40 percent of sales but generated more than 70 percent of sales growth. This significant differential, the report noted, demonstrates that selling healthier items is good business. Hank Cardello, the author of the report, said he hoped these findings would inspire companies to continue making better foods.

So what are all these salubrious offerings? The roster includes Baked Lays, Diet Pepsi, vitaminwater, Goldfish, Diet Snapple, Keebler Right Bites cookies, Rice Krispies, Spam Light, and Egg Beaters. Oh, and Go-Gurt. Had it been launched as a product, my low-calorie, fiber-enhanced biscotti surely would have made the list. The problem is that most nutrition experts would not consider these foods healthy; many of them wouldn't even pass the *healthy* sniff test with the average person. Yet stunningly—and revealingly—food companies thought the criteria was too stringent, according to Cardello.

This wasn't the first time the food industry had elasticized the concept of "healthy." In 2011, the Prevention Institute, a public-health nonprofit in Oakland, CA, looked at fifty-eight products with on-package labeling identifying them as healthy and concluded that 84 percent of them didn't meet very basic nutrition standards. And of course, there was that now-defunct, industry-created Smart Choices program, the one that dug its own grave by awarding check marks to Froot Loops and Apple Jacks.

Cardello acknowledges that his report didn't define "better for you" products according to what he calls "tight nutritional standards." This, he explains, was intentional—necessary to "jump-start a mass reduction in the number of calories sold." Most items on Cardello's A-list were there by virtue of their reduced or minimal levels of calories, salt, sugar, and fat. A sixty-year old MBA with a wide, toothy grin, Cardello is an easy guy to like. He's smart and upbeat, and when you talk to him, you get the sense he really wants to do something positive. After spending three decades in the food industry, working at Coca-Cola, General Mills, and Cadbury-Schweppes, he wrote a book called *Stuffed: An Insider's Look at Who's (Really) Making America Fat*. It helped launch his second career as an obesity and food-policy pundit.

Cardello believes obesity is health problem number one, and that food companies are well-equipped to tackle it because they're able to remove two big sources of calories from their products—fat and sugar—although he recognizes that doing so is never easy. Cardello believes that food companies should be encouraged in this endeavor, not coerced or bullied through regulation or government policies. "We've got to incentivize them to move in the right direction in a way that's also good for their business models. It's the carrot, not the stick," he told me.

Except it's not a carrot. More likely it's a vitamin A– and fiber-enhanced, sodium-reduced, baked, multigrain, garden-vegetable snack chip—a product that's healthy in the same way that buying a twenty-mile-per-gallon Escalade Hybrid is going to help wean us off foreign oil. Product reformulations often look good without actually accomplishing anything. The more fat and sugar you take out, the more they're replaced by other cheap, empty ingredients, whether zero-calorie sweeteners, starches, gums, or taste-modification molecules. And just because something is low in calories, this doesn't mean it's healthy, just as taking out artificial ingredients, while commendable, doesn't make a food good for us, only less bad.

Instead of encouraging truly healthier eating, all these remodeled, less-bad products can have the effect of keeping us tethered to a merry-go-round of confusing choices. I thought about this not long after my visit to Penford. Normally when I order biscottis with my coffee, I'm under no illusion that I'm eating something healthy. It's better than a cinnamon roll, but still pretty much crap as far as nutrition goes. I realize this, so maybe I balance it out by eating a salad for lunch. But if my biscotti contains an apple tree's worth of fiber, perhaps I persuade myself into thinking of this sweet carb concoction as somehow maybe a little bit good for me. Then suddenly I find myself eating biscotti more regularly or deciding I can get pizza for lunch with ice cream for dessert, ultimately making the biscotti more harmful than helpful. Eating healthy, at least for many women, is bound up in the stories we tell ourselves.

The official name for this semi-willful dietary deception is the Snackwell Effect, named after Kraft's brand of "healthier" cookies, introduced in the early nineties at the zenith of low-fat hysteria. The cookies—which had only 55 calories each and much of the fat replaced by emulsifiers, starches, and gums—were marketed as an indulgence you didn't have to feel guilty about. Eager for a hall pass on food guilt, shoppers literally ate it up. A store manager in Dallas reported he couldn't keep one particular variety on shelves: the Devil's Food Cookie Cakes. Customers were loading up on them by the case. Either confused about the cookies' true health status or willfully disregarding what they knew, many people gobbled up more cookies and more calories than they would have otherwise, obliterating any possible health benefit. Plus, they got heaping doses of white flour and sugar, which are probably more responsible for weight gain than fat is anyway.

The Growth Game

The thing about major food manufacturers that you hardly ever hear is that they do already sell bona fide healthy products. Just not very many of them. Kraft's Planters brand and Pepsi's Frito-Lay division make bags and tins of nuts; ConAgra sells David's sunflower seeds; Smucker's makes peanut butter that's nothing more than peanuts and salt; General Mills's Cascadian Farms brand offers an array of frozen fruits and vegetables, as does Pinnacle Foods's Bird's Eye brand. PepsiCo owns a stake in a coconut water company, sells the unsweetened oats that Quaker has been rolling out for 112 years, and has newer Real Medleys oatmeal cups with dried fruits and nuts; Tyson sells plain old chicken breasts; Dannon makes tubs of unsweetened yogurt; McDonald's pioneered the bagging of apple slices, unfortunately coupling them with caramel sauce; Wendy's has had simple baked potatoes on the menu for eons (though they're "baked" in a microwave). I could go on, though not for much longer.

These products—an ideal marriage of convenience and minimal processing—tend to get overlooked because they account for just a sliver of industry sales and are often less profitable than more exciting and complicated fare. Simple items like cheese, frozen vegetables, and chicken breasts have gross margins ranging from 15 percent to 30 percent. Breakfast cereal and snack chips, on the other hand, command margins up to 70 percent; soda and sports drinks offer a ridiculous 90 percent. This is why you see a constant barrage of ads for Gatorade and nothing for frozen blueberries.

Product profitability is as much a necessary consideration for food companies as how their products taste. Nearly all large packaged-food manufacturers are publicly traded corporations and have to show continual growth not just every year, but every quarter. Despite what may be the genuine good intentions of the people who work there, these companies are simply not set up to sell very many authentically healthy foods, and their investors really wouldn't want them to. Trained to look at companies by the numbers they generate, financial analysts and large stockholders are mostly indifferent to whether a company's foods are healthy, regardless of who's defining "healthy." You won't find anyone on Wall Street referring to hundreds of millions of dollars of snack chips or soda sales as a liability.

Executives at PepsiCo bumped up against this reality not long ago. Over the past decade, PepsiCo has evolved into the U.S. food company with the greatest level of purported concern for nutrition and health. It spends more time talking about the subject than others and has made a series of unconventional, high-profile hires to help take the company in a new direction. It brought on a public health hero from the World Health Organization, a former endo-crinologist at the Mayo Clinic, and an official from the Centers for Disease Control. In 2011, the former WHO official tried to explain his decision to join a company so strongly rooted in soda. "While we are not likely to become a fresh fruit and vegetable company, we have made public commitments to increase the use of fruits, vegetables, nuts and whole grains in our products," Derek Yach, Pepsi's senior vice president of global health and agriculture, told public health lawyer and blogger Michele Simon. "A major chal-lenge involves ensuring that we do so in ways that maximize the full nutritional equivalence of whole foods in our future products."

Simon wasn't sure what "the full nutritional equivalence of whole foods" meant, but it was nonetheless encouraging to hear a food executive engage in serious conversation on the issue. The burst of corporate introspection had been driven largely by the galvanic presence of Indra Nooyi, a CEO unlike any PepsiCo had had before. The first woman and the first Indian to lead the com-pany, she took on nutrition as one of her defining issues, seeking to recast what a food company could be. She acquired a majority stake in a Russian manufacturer of fermented dairy beverages and charged the Frito-Lay snack division with an ambitious agenda of removing all artificial ingredients from their products, start-ing with Sun Chips, Lay's, and Tostitos. She talked about healthy "drinkable snacks" and "snackable drinks."

And then investors and Pepsi bottlers freaked out. In March 2011, the trade journal *Beverage Digest* reported that Pepsi-Cola, the flagship brand, had slipped to third most popular beverage product in the United States, with Diet Coke usurping its number two spot. Regular Coke was still the leader. This was a major upset in the beverage world, and Ann Gurkin, an analyst with invest-ment firm Davenport & Co., knew exactly why it had happened.

"A good amount of the weakness stems from the management's increased focus on its better-for-you portfolio. They took support and focus away from their core beverages—Pepsi-Cola, Diet Pepsi—and spent more time and more effort enhancing juices, waters, sports drinks," she lamented to *Business Week*. Pat Weinstein, the owner of a Pepsi bottler in Seattle, put an even finer point on it. "Is she [Indra Nooyi] ashamed of selling carbonated sugar water?" he asked.

Nooyi quickly responded by redistributing some $500 million ad dollars into soda marketing, including a $60 million sponsorship deal with Fox's *X Factor*, an answer to Coca-Cola's decade-long sponsorship of *American Idol*. She hustled Pepsi back into the Super Bowl after declining to run ads in 2010 and 2011, the first time in twenty-three years. And she appointed someone new to head the sagging beverage division, a thirty-year company veteran for whom the investor unrest had served up a valuable lesson. Shortly after he was promoted, Albert Carey told the *New York Post* about the new approach he'd be taking: "The one thing I did realize in the last twelve months is that you can't just try to promote healthy brands alone—you have to also go in behind indulgent brands," he said, speaking about a successful Ruffles Hot Wings marketing campaign. "I wouldn't have taken this position if I was embarrassed to sell all these products."

Carey is right. At the end of the day, PepsiCo is a company that sells Ruffles Hot Wings, Doritos Jacked, Cheetos, Lays, Chewy Dipps granola bars, Pepsi-Cola, and Mountain Dew—products that aren't the least bit healthy but are profitable and popular, thanks in no small part to heavy marketing assaults over the years. And if there's one thing Nooyi's cash influx to soda demonstrates, it's that consumer eating habits are partly a function of how much money companies spend influencing them. As of the fall of 2012,

PepsiCo's North American soda sales were showing signs of initial turnaround. The moral of the story was clear: for a company like PepsiCo to have downplayed some of its biggest sellers, whether intentional or not, was nothing short of financial recklessness.

The High Bar

While most large packaged-food companies are traded on stock markets, there is one that isn't. And not only is Mars, the candy and chocolate purveyor, not public, it's private in the best possible way. Its stockholders, who have been members of the Mars family for the past 102 years, tend to be patient and long-term, willing to give the company's management a generous leash. It helps that Mars executives don't have to go before a band of number-hungry financial analysts every three months to justify their spending outlays and profit returns. If the company's management is so inclined, they can spend millions of dollars on original health research and out-of-the-ordinary product development that may not provide any real returns for upwards of two decades.

This is exactly what the company has done. Starting some twenty years ago, Mars's scientists quietly set out to understand everything there is to know about the flavanol phytochemicals that reside in the cocoa plants used to make chocolate. It had long been suspected that chocolate, despite its copious quantities of fat and sugar, might contain health-promoting compounds, at least in dark varieties. The Kuna Indians in the San Blas Islands of Panama, for instance, have low blood pressure levels that don't rise with age—owing, scientists think, to their consumption of a drink made from cocoa beans.

Over the years, Mars's researchers, both on their own and in collaboration with academic scientists, published more than one hundred and forty papers on these natural compounds. They charted the chemical composition of flavanols, the ways they're metabolized by the body, and what benefits they offer. The conclusion: flavanols help the body access its own natural pool of nitric oxide, allowing blood vessels to widen so that nutrients and oxygen in the blood can flow freely, improving the circulation of these nutrients to the body's tissues. Mars conducted studies in humans showing that when the right amounts of flavanols were consumed, oxygen and nutrients got to where they needed to go in the body.

Then the company's food scientists started looking for ways to preserve these useful substances during chocolate production without sacrificing flavor, serving up a rare example of technological know-how in pursuit of nutrition preservation. Flavanols have an intensely bitter taste, which is part of the reason they're eradicated during processing. They're damaged not only during the fermenting and roasting of cocoa beans but also during the chemical alkalization process that makes chocolate taste smoother. Mars's food scientists spent several years trying to figure this out, ultimately settling on a combination of techniques for preserving flavanols—using the right beans, limiting fermentation, skipping alkalization, and carefully controlling bean roasting. The company dubbed the new method CocoPro and tried to market it in 2006—with disastrous results.

Consumers simply couldn't adjust to the notion of "healthy" versions of dark and milk chocolate bars from a company selling vast quantities of M&Ms, Snickers, and Milky Ways. Those inclined to buy healthy products didn't want them from a candy purveyor, and those inclined to buy Mars products weren't interested in something that purported to be healthy. The concept, it seemed, was a bit ahead of its time. Plus, the new chocolate was frequently sold in the "special needs" section of grocery stores, and no one knew where to find it. For a company not accustomed to marketing healthy food, it was a humbling lesson. "We recognize that chocolate can be part of a healthy diet, but in and of itself is not a health food," Catherine Kwik-Uribe, head of research at Mars Botanical, said in a phone conversation.

So the company took a different approach and in 2010 started selling another product, one that's a telling illustration of what's really at stake in the creation of healthy processed food.

One afternoon I headed over to King Soopers to look for it. Mars happened to be testing the item, a type of bar called Goodness Knows, in stores in Denver and Boulder. I found them in the section of the supermarket that sells only wholesome things—you know this because in those three aisles, they've laid down walnut-colored wood flooring instead of the standard high-gloss linoleum. These bars, which Mars describes as "luscious snack squares," were nestled between boxes of Health Valley Organic chocolate multigrain toaster tarts and Late July vanilla bean sandwich cookies. The company advertised in small type at the bottom of the package that the bars "support circulation of nutrients." The point was visually illustrated by a series of long arrows dancing around a human figure.

When I tasted several varieties of the bars later at home, I discovered that I was eating all the items pictured on the package, a surprising occurrence. The almonds, dried cranberries, hazelnuts, oats, chocolate, and dried raspberries were there in the bars, where I could see them. More often than not, those scrumptious-looking fruits and vegetables on food packages are nowhere to be found once you pry the pack open. No one actually expects Garden Salsa Sun Chips to contain anything resembling a tomato or pepper, or for there to be whole blueberries in a box of blueberry breakfast bars.

Although it was a sweet snack item, not a meal, Goodness Knows was more food, less food product. About half of the ingredients were whole or minimally processed. And thanks to the oats, nuts, and dried fruit, it required chewing. Sugar, fat, and calorie levels were reasonable, and sodium was negligible. The dark chocolate enrobing the bottom of the bar promised 200 milligrams of cocoa flavanols per serving—a scientifically researched and measured, naturally occurring phytochemical content you'd be hard pressed to find elsewhere in the supermarket's middle aisles. And for all these reasons, it was more expensive than other bars. When I bought them, they were 40 percent more expensive than Clif Bars and 20 percent pricier than LUNA bars.

Goodness Knows was in product development for eighteen months and then in a ploddingly slow test in just two cities for two years. As Catherine Kwik-Uribe explained it, Mars wanted to make sure Goodness Knows didn't go through the standard bar extrusion process or high-temperature baking. "It was really important that we retain the integrity not just of the flavanols but of the all the original ingredients," she said. "If you look at some of the other bars in the category, it may call itself dark chocolate raspberry infusion, but you look at that chunk in the bar and you're not really sure what it is." Kwik-Uribe said sales have been good, in accordance with expectations, and that Mars has recently expanded availability to Seattle and Portland. Then maybe, sometime down the road and over the rainbow, there will be profits.

Mars's healthy snack adventure gobbled up over two decades and close to $100 million. It turns out that when you apply a stringent definition to the word *healthy*, healthy processed food can be quite difficult and expensive to create. And this is why we really shouldn't blame food manufacturers for coming up with Baked Lays and FiberOne 90-Calorie Brownies instead of products with whole ingredients and naturally occurring phytochemicals. The executives that run food companies aren't bad, lazy, or unimaginative people. They may succumb to financially motivated wishful thinking about what constitutes healthy food, but that's because they're doing what they're good at and what they're rewarded for: industrially processing foods to make them profitable, and then marketing the hell out of them. Often this requires them to make products that look healthy but really aren't.

It's not that large packaged-food companies can't have any impact. It's that their contribution to the overall problem is always going to be miniscule. Pepsi, Kraft, Kellogg's, ConAgra, and General Mills won't be the ones to improve our diets and fix our health problems, and we should stop asking them to.

11

Sit at Home and Chew

The destiny of nations depends on how they nourish themselves.

—Jean Anthelme Brillat-Savarin

In 1983, a commercial for Mrs. Paul's frozen fish dinners appeared on television. It featured a woman wearing a pair of oversized glasses, a pearl necklace, and a high-necked, eighties-style purple dress with multiple bursts of ruffles. She stood in her kitchen looking defeated and forlorn. In one hand, she clutched a sharp knife and in the other, an enormous fish—head, gills, and all. "While most everyone loves seafood, not everyone knows exactly what to do with it," a voice-over said. "Introducing Mrs. Paul's light and natural entrees—seven entrees low in calories and micro-wavable." The commercial ended with an image of a man hunting down a loose lobster in his kitchen. The voice-over: "For people who love seafood, but don't love to cook it."

I saw this ad and others like it at the Paley Center, a museum in Midtown Manhattan where I spent an afternoon watching food commercials from the sixties, seventies, and eighties. The center houses an impressive collection of old radio and television ads and programs, and if you spend enough time there, you can start to feel as though you've launched yourself into a time capsule. I relived the days when OJ Simpson ran through airports, when Cher hawked perfume and Frank Perdue personally vouched for his chickens, something he did with disarming charm from 1971 to 1994. Frank, I noticed, didn't change that much over the years, but his products did. Ads for raw chickens and chicken parts in the seventies gave way to appeals in the eighties for precooked breaded cutlets and nuggets that were useful for fooling people into thinking you'd cooked.

I was interested in the messages food companies have delivered about the basic human art of cooking over the years, and the discourse turned out to be remarkably consistent. Among the dozens of commercials I watched at the Paley Center, all of them that had something to say about cooking were tailored around a simple idea: messing around in the kitchen is a cumbersome, time-consuming chore. Why bother with such labor when drive-throughs can be visited and trays deposited into microwaves?

Along with the Mrs. Paul's ad, there was this memorable one from 1988 for Rich's Southern Barbeque. In it, Ruby, a waitress at the Pork-O-Rama, gets home from a long shift and proceeds to do "diddly squat." She sports a bright-orange waitress uniform and fuzzy pink slippers. "I keep these new Rich's Southern Barbeque sandwiches on hand at all times," she says, closing the freezer and popping open the microwave. "You've got your hickory smoked pork and your chicken. You don't have to buy buns. Don't have to buy sauce." Sandwich in hand, Ruby sashays into the living room and sits down in her reclining chair, tossing up her pink feet. "Just sit at home and chew. I do love the leisure life."

In the fifties, Jell-O instant pudding took the opposite tack and portrayed women as having no leisure time at all (perhaps because frozen pork sandwiches hadn't been invented yet). In an animated commercial, a housewife walks on a treadmill with a broom in hand. All at once, the phone rings, a hand knocks on the door, a baby screams, and empty pots and pans shake with anticipation. "Dinnertime, oh dinnertime. Too late to make dessert," a creepy female voice-over laments, followed by one of those magisterial fifties male voices informing us that Jell-O instant pudding is "ready-to-eat, creamy, nourishing, so delicious, and so quick." A "No Cooking" banner on the box hammers home the point.

Mrs. Ruffles, Ruby, and Treadmill Mom belong to a genre of food advertising going back to the 1940s when manufacturers shifted from an emphasis on food's technological modernity to an appeal based on the virtues of not having to cook. In her book *Something From the Oven: Reinventing Dinner in 1950s America,* Laura Shapiro describes the food industry's post–World War II dreams of "a day when virtually all contact between the cook and the raw makings of dinner would be obsolete." As emphatic as they were, such efforts to introduce American housewives to a new generation of prepacked meals weren't immediately successful. "Straight through the fifties women kept cooking. Willingly, haphazardly, in a lousy mood or a panic, sometimes with enjoyment and sometimes hardly noticing the food, they cooked," Shapiro notes. It wasn't until at least a decade later, when women began returning to the workforce, that the food industry's exhortations to abandon the kitchen actually started working.

And they kept working. Over the last seven decades, home cooking in America has plummeted. In 1927—the pre-TV dinner era—the average woman spent an unimaginable five to six hours a day preparing meals for her family. By the fifties, the food industry claimed that a housewife relying on convenience foods could fix her family's meals in an hour and a half or less, which is still an eternity by today's standards. These days American home cooks do, on average, barely thirty minutes of cooking a day, the shortest food prep time among thirty-four other developed countries, according to a 2011 survey by the OECD group of nations. The Chicago research firm NPD tracks the decline in terms of percent of dinner entrees made from scratch; it's dropped from 72 percent in 1984 to 59 percent in 2010.

Replacing home-cooked dinners are fast-food and frozen items. If it seems as though supermarket freezer aisles keep getting bigger, that's because they are. To accommodate hordes of new products, frozen aisles have been supersized to an average total of 400 feet, a length that would span more than a football field and is double what it was in 1990. And those shiny, glass-walled aisles could get bigger still. Sales of frozen foods and beverages in all U.S. retail outlets are expected to jump from $56 billion in 2010 to $70 billion by 2015, a big leap for the slow-growing food business.

The chief factor driving this decline in home cooking is, of course, the migration of American women into the workforce, starting in the mid-sixties and continuing to the point where 40 percent of women aged sixteen and over now have full-time jobs (another 14 percent have part-time jobs). Much as the washing machine and dishwasher have, the wide array of instant food options unshackled women from the burden of nonstop manual labor, allowing them to pursue ambitions beyond the home. Yet even in two-income households, studies show that women still end up doing the lion's share of housework. And women do nearly all of it in single-parent households, which are overwhelmingly female and have increased by 74 percent since 1980 (versus 35 percent population growth).

The Grocery Manufacturers Association, the packaged-food lobbying group, has quantified time savings by comparing the average amount of time spent doing cooking and other food prep in 1961 versus today. Thanks to both the packaged-food and restaurant industries, American families save eighteen days a year, or one and half days per month, and a little over an hour a day. "As a mother, I know that it has never been easier for parents to fix nutritious, convenient and affordable family meals," the group's president, Pamela Bailey, wrote in a *New York Times* letter to the editor. "The benefits of family meals are clear, and food and beverage manufacturers will continue to make it easier for families to dine together."

Whether packaged and fast food has yielded more family time (or more TV and Facebook time) is debatable. But the unassailable fact is that even when women don't have jobs, they often cook less than their stay-at-home mothers or grandmothers did. Sometimes it's because we're tired, and it's easier not to worry about what ingredients may or may not be in the fridge. Other times it's unfamiliarity, fear, and sheer force of habit. Kitchen illiteracy gets passed down from previous generations. If we don't learn cooking skills from our parents or other relatives, how are we supposed to approach the endeavor with any confidence or enthusiasm—especially when characters like Ruby, Mrs. Ruffles, and so many others have assured us we really don't have to bother? According to a Pew Research Center survey several years ago, just 34 percent of Americans say they enjoy cooking "a great deal," while another 26 percent say they enjoy it "a fair amount."

Yet some Americans do manage, despite hectic schedules and an absence of culinary mentors, to prepare food from fresh ingredients in a way that doesn't require hours of sweat or an advanced culinary degree. In Sacramento, I met a woman who decided to do what so many in the food industry say Americans don't want to do. She orchestrated a full overhaul of her family's diet and started eating as if health mattered. After years of making quick, thoughtless choices, she no longer wanted to "just sit at home and chew."

Coming out of the Fog

Darcy Struckmeier flips open her MacBook to show me some photos of her son she took about two years ago. Cameron is a tall, lanky thirteen-year-old, and in the pictures he's a tall, lanky eleven-year-old. In one, you see the ocean in the background, and in another he's posing with a friend and cracking an awkward, hardened half-grin, looking as if someone made him smile for the camera. The rest of the photos show a boy with a flat expression and heavy, glassy eyes that don't communicate sadness so much as a confused indifference. Under his eyes are hollowed, dark circles. You wouldn't think much of this collection of snapshots unless you happened to know Cameron today. "It's like he's a different child," Darcy marveled, staring at her computer. "There was something there before preventing him from being happy."

Darcy and her husband Shawn have four children—Cameron, his two sisters, Taylor, age twelve, and Emma, seven, and twenty-two-month-old Greyson. They live on the south side of Sacramento in a neighborhood that sits between a rough, gang-infested patch of suburbia and the rural edge of one of California's largest cattle-ranching areas. Their home is a modest two-story unit with a little backyard that sits in a quiet new development. It's a middle-class area that, like most, has its share of good and bad. Along with the endless number of balmy sunny days and no shortage of wonderful neighbors, there are shootings, break-ins, and police helicopters thumping overhead. Darcy won't let any of her kids, even thirteen-year-old Cameron, wander over to the park across the street on their own, thinking it too big a risk. A few years ago, she decided to home school her kids. Many lessons take place when Greyson is napping. Darcy is thankful that he's a good sleeper.

When the Struckmeiers moved here six years ago so that Shawn could take a job as the musical and creative arts director of a nearby nondenominational Christian church, Cameron was seven and already having behavioral problems. He'd always been an exceptionally stubborn and willful child, and not just about certain things or only when he got tired. It was all the time; he fought his parents on everything, especially food. Cameron refused to eat anything that wasn't crackers, milk, pasta, or fruit. At times, he would throw up at the table from foods he didn't like, and he often complained that his stomach hurt. As he got older, his angst intensified. "He never really seemed happy or joyful like most kids do. It was hard to get him to smile," Darcy recalled, as we sat on a navy-blue sectional couch in their small living room. When he turned eleven, Cameron was able to articulate more about how he felt. He described his insides as being stretched like a rubber band, as tight as possible and about to break. He felt angry all the time, he said, but he didn't know why.

The Struckmeiers' doctor suggested Cameron start taking ADHD medications, which he did for several months. When they seemed to have only negative effects, Darcy took him off them. She then had him tested for food allergies, but the verdict was that he had none. Darcy and Shawn wondered if their son might have Asperger's; they took him to the Mind Institute at the University of California at Davis, a noted center for neurodevelopmental disorders. But Cameron's behavior was deemed mild compared to the problems of most kids who show up there, and he and his mom returned home with no solutions and no diagnosis other than that Cameron had sensory processing issues, which they already knew. Out of options, Darcy and Shawn resigned themselves to continue taking it day by day, giving Cameron all the love and support they always had. Then one night Darcy, who is thirty-six and has the bright smile and unflappable enthusiasm of the cheerleader she was in high school, stumbled on a Web site that would turn her life upside down. It was urging people to take a ten-day challenge to give up all processed food.

As someone who's always struggled with her weight, Darcy had been searching for a sane, long-term approach to eating healthy. She was done with strict diets and prescription diet pills, all of which had helped her lose heaps of weight, only to pack it on again months later. The idea of a zero-processed-food challenge appealed to her all-or-nothing personality, and its emphasis on fresh, healthy foods instead of low-fat, low-carb, or calorie-restricted nonsense clicked with her. She decided to give it a try. Called 100DaysofRealFood.com, the site is run by a mom named Lisa Leake in North Carolina, who took the challenge herself several years ago for a hundred days. Darcy bookmarked info on what was and wasn't processed food and she looked over various posts about food ingredients. She printed out a dozen recipes that seemed yummy and easy. Darcy wasn't intimidated by cooking— she had learned her way around the kitchen from her mom—but she never did it very much. About four or five times a week, the family went to Chick-fil-A, McDonald's, or Jack in the Box. At home, they ate frozen chicken nuggets, Lean Cuisines, Papa John's pizza, Pop-Tarts, and Nacho Cheese Doritos.

Darcy figured she'd do the challenge alone, as she always had with her diets, but when she told Shawn about it, he said he wanted to join her. "I knew we ate really crappy," he said when I asked him about it during my visit. "We went to fast food all the time and it didn't feel right to me. I didn't grow up eating that stuff."

Shawn and Darcy presented the idea to the kids to see if they wanted to go along for the ride, figuring they wouldn't, especially Cameron. They told them that for ten days, the family wouldn't be eating fast food or going to any other restaurant. Everything would be prepared at home from fresh ingredients from the grocery store. There would be no candy, and snacks would be things like carrots and hummus, instead of chips and Go-Gurt. When it was all over, after the ten days, they could go to a local cupcake shop called Icing on the Cupcake. To Darcy and Shawn's surprise, all three kids said they wanted to do it. Greyson, of course, was still too young to have any say in the matter. Darcy figured they probably didn't realize what they were signing up for. Or they were just fixated on the cupcake.

During the first few days, Darcy, who's always been in charge of the family's meals, tried to make the new food fun and exciting. One night, she made chicken stir-fry with brown rice, pineapples, and apricot spread, and the next, spaghetti and meatballs. For dessert, she made whole-wheat crepes with chocolate sauce and maple syrup, a recipe from Lisa's Web site. The next night she cut up pears, drizzled honey over them, topped them with raspberries and placed them in cute little dishes. "Everybody was so impressed. The kids were like, 'Mom, this is so fancy,' though it really wasn't," said Darcy.

On the morning of day seven, something unexpected happened. Cameron came downstairs and told his mom he felt different. "I feel like I've been lifted from a fog," he said. "I feel like I can think better." Darcy was stunned; her son had never said anything like this before.

Then about a week later, after the ten days were over and the kids had had their victory cupcakes, she realized something really had changed with Cameron. She and the kids were at Target, shopping for clothes, when she suddenly caught her son's eye. She could tell he was smiling. "There you are," she thought to herself, realizing at that moment that she'd only ever seen the blank outlines of her son. He'd never been fully drawn in.

There were other welcome changes too. Emma had always had constipation, and on day five of the challenge, she came downstairs with a worried look on her face. "Mom, I have diarrhea," she announced. But when Darcy went upstairs with her, she realized her daughter had simply had a normal bowel movement, something that hadn't happened in such a long time, the seven-year old didn't recognize it. Shawn's health, too, improved. His persistent postmeal heartburn nearly vanished; the Tums container now sat undisturbed in the medicine cabinet. And he'd lost eighteen pounds without trying, which both delighted Darcy and drove her crazy. "It's ridiculous how men can just melt away pounds," she sighed. "I'm out there running five miles—and nothing."

Oddly enough, it was Darcy—and Taylor, who didn't haven't any health issues—who reaped the fewest personal benefits from the Struckmeiers' eating adventure, which everyone agreed deserved to be upgraded from a ten-day challenge to a new form of normal. When I visited the family in September, six months after the March challenge, Darcy had yet to lose the thirty pounds she was aiming for, although she was awed at the monumental improvement in her energy levels. She said she can now notice when she's full and needs to stop eating and when she's truly hungry, not just having a sugar crash. "It used to be that if I didn't eat a Krispy Kreme right away I was going to gnaw my arm off," she said.

It's easy to see how cutting out most sugar and white flour could stabilize Darcy's blood sugar levels; how increased amounts of fiber from fruits, vegetables, and whole grains would have helped Emma's constipation; and how the absence of greasy fast food might have alleviated Shawn's heartburn. But it wasn't clear how to account for Cameron's transformation. I asked Darcy what she thought it was about their previous processed-food-rich diet that had either caused or exacerbated Cameron's problems. She said she wasn't really sure. "It could be the sugar, the food dyes or all the artificial ingredients," she suggested. "Or a cocktail of all three." Most doctors and scientists, too, would be at a loss to pinpoint what exactly might be going on. Barely any food additives have undergone testing for links to behavioral disorders.

But Darcy is certain it's the food. This became quite obvious when the family went on vacation in June to Paradise, California, a small town near the Plumas National Forest. They stayed in a hotel without kitchen facilities, and the town didn't have many options other than fast food. Darcy had packed some rations from home, but when they ran out everyone started eating pizza and fast-food burgers and fries for dinner and pastries and donuts in the morning. The kids went to Carvell and Baskin Robbins. Four days into the vacation, Cameron started having angry outbursts and refused to leave the hotel room. When his parents sat down to talk to him, he said he felt awful and that his stomach hurt. "I just want to punch everybody," he said.

When they got home, Cameron went back to eating his old, new diet—Shredded Wheat or yogurt and fruit for breakfast, grilled cheese sandwiches on whole-wheat bread for lunch, carrots, snap peas, and nuts for snacks, and whole-wheat pasta for dinner. Four days later, his anger, his stomachache, his antisocial behavior—was all gone again. "I think that was a real learning point for him too," Darcy said. "He realized how all that food made him feel and now he works on trying to avoid it."

Darcy still marvels at the situation. "It's hard to believe the food that you've been feeding your kid his entire life is what was making him like that. It's the craziest thing." And she's still trying to adjust to her own metamorphosis—from someone who used to consume Jack in the Box tacos and a forty-four-ounce Diet Coke for breakfast (only 360 calories!) to a person who knows what kale and tahini are. "I'm not one of those born healthy eaters," she wrote on her blog. "Quite the contrary. I'm a convert, through and through. I just have to say from the bottom of my heart that if I can do it, truly anyone can."

On the evening I joined them for dinner, Darcy had laid out the fixings for chicken tacos on the kitchen counter. It had taken her about thirty minutes to put together, not counting several chicken breasts she had thrown into the crock pot along with some Trader Joe's salsa earlier in the day. There was shredded cheese, corn (from a bag and frozen), sour cream, and shredded cabbage. A basic salad of romaine lettuce and bell peppers was in a large bowl next to a vinaigrette Darcy had mixed together. The only thing that seemed elaborate about the meal was the homemade whole-wheat tortillas she rolled out on her counter and cooked in a fry pan. She insisted they were easy.

After Shawn said grace, we all started eating. Except for Cameron. He was at the stove putting together his taco, which was more of a quesadilla. He omitted the chicken and passed on the salad. Cameron, who has close-cropped blond hair and the nervous, self-conscious gaze of an exceptionally bright teenager, said that although he's still very particular, he's trying to eat more types of food. "I almost like chicken," he said. When you're able to catch his glance, his eyes are bright and his face is handsome. There's no evidence of the blank, glassy look I saw in the photos.

At first, figuring out all the food prep was daunting, but Darcy said she now has a system. She plans out the week's meals and snacks ahead of time and does chopping, washing, and other necessities immediately after each weekly shopping trip, so that much of what's in the fridge is ready to go into a pot or the oven. Most nights, dinners are simple, inexpensive affairs—baked potatoes with veggies and cheese; whole-grain pasta with sugar snap peas, corn, carrots, balsamic vinegar, and fresh-grated cheese mixed in. Often times it's leftovers from several nights earlier.

The biggest challenges arise beyond the home. The Struckmeiers' church serves candy, donuts, cookies, and Cheez-Its for the kids in morning Sunday school. And nearly everyone they know eats fast food regularly. Darcy says she tries not to get too obsessive; if they go over to someone's house for dinner and the meal is hot dogs and white pasta salad, they eat it. If family members invite them to dinner at the Old Spaghetti Factory, they go. And sometimes the kids are good about turning down sweets; at other times, they're not.

Making sure everyone is at home for dinner also isn't easy, especially since so many kids' activities seem to be scheduled during dinnertime, or what used to be dinnertime. Shawn works late three nights a week, and sometimes one of the kids has to be somewhere, though Shawn and Darcy actively try to avoid overscheduling Cameron, Taylor, and Emma in too many activities.

Darcy and Shawn have a strict food budget of $800 a month, and it can be a struggle to make it last, in part because they've decided to buy as much organic food as possible, due to its lower pesticide residues and the absence of genetic modification. To help make ends meet, they opted to give up cable a few months ago, which Darcy explained as a choice based on values. I asked Darcy whether she thought they would eventually lose enthusiasm for their new path and revert back to the ease of chicken nuggets and drive-throughs. "There's no way," she said, without hesitating. "Everything we're eating now tastes so much better, and we all feel so much better, so it's completely worth the increased time and cost to us. I get a lot of people saying that healthy eating is some weird, bad-tasting, nasty food. I thought that, too. Instead I found immediately that real food tastes better. I don't mean slightly better, I mean my kids are asking for seconds and thirds better. Now fast food is what tastes weird and disgusting."

I spoke with a middle-aged woman named Laura in New Hampshire, who was also amazed at how switching from sugary, refined, prefabbed meals to fresh ones had transformed her life. She'd overhauled her diet after hearing about a workplace program offering health coaches, cooking advice, and nutritious foods served at her corporate cafeteria. Like Darcy embarking on the ten-day-challenge, Laura needed a push and a little support to do it. The more people I talked to, the clearer it became that this is something we all need. In truth, it's hard to see how America gets off its suicidal diet without interventions emphasizing the importance of real food.

Some such programs are already out there. Back in Colorado, I attended a series of cooking classes run by a Washington, DC–based nonprofit organization called Cooking Matters. Currently offered for free in thirty states and designed with lower-income communities in mind, it's the largest national program teaching people how to cook. You can learn cooking at places like Sur La

Table and various cooking institutes and colleges, but this route isn't cheap. When I checked at Sur La Table, twelve hours of basic classes cost $420.

On the August evening when I showed up for the Cooking Matters class held in the kitchen of Longmont, Colorado's First United Methodist Church, Jean Bowen was showing a class of seven women and two men how to assemble a barley jambalaya and an apple and peach salad. The plan had been to teach the students how to make the apple salad found in the recipe book they all get with their class materials, and the jambalaya wasn't supposed to include eggplant or chard. But it was the height of summer, and Bowen had made some spontaneous selections at the farmer's market. "I thought these looked amazing," she said clutching a small bag of peaches. "And you can't believe how great they smell." She passed them around.

The students stood around shiny steel cooking tables staring blankly. Everyone had an onion, garlic clove, or celery stalk staring up at them from a cutting board. Scattered around the tables were clumps of green chard leaves, green peppers, eggplants, some chicken breasts, apples, and bundles of beets with the leaves still on.

"Here's the thing," Bowen said, digging her knife into an onion. "If you cut an onion just right, you can do it quickly before it makes your eyes tear up. Who starts crying when they cut an onion?" A few hands went up. Bowen sliced into the onion horizontally, then vertically, then straight down from the top until it dissolved into a pile of tiny cubes. She scraped the pieces into a large pot waiting on the stove and then instructed everyone to start chopping. The room hummed with chatter and chopping.

Like all the other trained chefs who teach Cooking Matters

classes, Bowen is a volunteer. This was her thirtieth class. She told me she'd studied cooking at the Art Institute of Colorado, eventually starting a catering company. When I spoke to her several weeks after the class, she explained that the food knowledge that students arrive with varies considerably. While some bring a deep appreciation for cooking and just want to learn some new recipes, others know some basic skills, like how to cook scrambled eggs and make a pot of rice. Still others have barely ever used a regular oven and have formed deeply codependent relationships with their microwaves. This group, as might be expected, is the one to get the biggest benefits from the classes. "There's such a big need for these basic kind of skills," Bowen said. "I've had people come up to me on the street and tell me they've lost twenty, thirty, or forty pounds mostly because of what they learned from my class. And there's an incredible sense of pride people get in being able to feed their families home-cooked meals."

In this respect, you might even say that the classes have the potential, as Michelle Obama might put it, to "help shape the health habits of an entire generation." Funded by WalMart and, somewhat discordantly, the ConAgra Foundation, the program is intended for those who receive food stamps and/or free or reduced-cost lunches for their kids. People of more ample means who wish to take classes, said Janet McClaughlin, Cooking Matters's national program director, may do so by serving as volunteers. Each of the six weekly sessions is organized around two practical recipes—much more Rachael Ray than Julia Child. There are no lessons on how to make homemade chicken stock, roast the perfect rack of lamb, or whip up mayonnaise from egg yolks. Boxed chicken stock, frozen vegetables, and canned tuna are part of the menu. For many people, such time-saving ingredients constitute advanced cooking. The spiral-bound recipe book that participants get at the start of the classes features dishes made with lots of vegetables, beans, chicken, fish, whole grains, spices, and moderate amounts of fat. After each class, students leave with a bag of ingredients that allow them to make the dishes at home.

The remarkable thing about all these recipes—besides how tasty they are, or at least the half a dozen I tried during the classes were—is how little they cost to make. Each meal costs no more than $10 for a family of four. This is roughly what you'd spend for two frozen pizzas from the supermarket. Or compare it to the $12 you'd pay for four crispy chicken sandwich value combo meals at Wendy's. Or the $16 you'd part with at McDonald's for four double cheeseburger meals. Although fast food and other processed options are often portrayed as affordable, the reality is that doing our own cooking is almost always cheaper, not to mention more nutritious and satisfying. You can see this cost differential in the frozen food aisles, where processed chicken is anywhere from 30 percent to 100 percent more expensive on a per-ounce basis than the raw stuff over in the meat department.

Eating well is in no way a luxury of the rich. We certainly can't all eat at fancy restaurants, but affordable nutritious food is more available today than it's ever been. Eggs, fresh meat, real cheese, plain yogurt, wholesome grains, half a dozen different kinds of beans and nuts, dozens of different fruits and vegetables—they're are all available pretty much any time we want them. The modern supermarket shimmers with a perplexing array of complex pseudo-foods, but it also holds the keys to the most nutritious and varied diet Americans have ever had access to. (Imagine our turn-of-the-twentieth-century ancestors contemplating the colorful bounty of supermarket salad and sushi bars.) The only trade-off (yes, there is always a trade-off, with real food as with processed food) is that we have to carve out a little more time and energy—

though not much more—to do some of our own food processing. As Barbara Kingsolver wrote in *Animal, Vegetable, Miracle*: "Cooking is the great divide between good and bad eating."

What we cook doesn't have to be grass-fed beef, five-year artisanal cheese, heirloom tomatoes, fresh-caught king salmon, or the self-produced food Kingsolver and her family spent a year nurturing. Buying organic, as the Struckmeier family does, is a terrific idea for reasons related to pesticide residues, farmworker safety, environmental sustainability, and perhaps enhanced levels of phytochemicals. Studies have shown that grass-fed beef and pasture-raised chicken and eggs have higher levels of omega 3 fats, vitamins A and E, and antioxidants like beta-carotene. But eating organic isn't necessary for optimal nutrition. Regular beef cooked into a vegetable stir-fry and set atop brown rice qualifies as a first-rate healthy meal.

The solution to all of our various health problems brought on by poor eating habits, after all, is in our own hands. Not those of the mega food companies. While there are clearly policy changes that would make the job of cleaning up our food a whole lot easier—a more equitable and health-minded distribution of farm subsidies, strict curbs on marketing to kids, food education in schools, for instance—the choice about what we feed ourselves and our children is ultimately ours.

A New Food Ratio

I don't mean to suggest we shouldn't ever eat processed food, or that the world doesn't need any food scientists. I have to confess that during the time I spent working on this book, my kids ate more precooked food than ever before. Being able to get dinner ready while you're sending e-mails or cleaning up a mess from the night before makes life feel a little easier. I still tried to seek out the least processed of the processed choices. But my feeling is that all the foods I wrote about in the book—bright-orange processed cheese, extruded and gun-puffed breakfast cereals, products loaded up with added nutrients, puffy breads with unpronounceable ingredients, things fried in unstable, hexane-extracted vegetable oils, food-esque soy protein, liquefiable chicken nuggets—they can all have their place—as long as that place is a small one.

One of the industry's popular cliches is "Everything in moderation," and this strikes me as useful advice on how we might approach the consumption of processed food. The food industry intends this expression to mean all of their various products consumed daily in reasonable quantities. A more meaningful interpretation of this edict is that all engineered food products go into one big, hulking category of what Cookie Monster would call "sometimes foods."

If the proportion of these items in the American diet today is 70 percent, perhaps a reasonable goal would be 30 percent, with the remainder coming from foods that have had a relatively direct and boring journey from the farm—foods that are, for the most part, grown not made. Such a rebalancing, though, requires a wholesale reevaluation of the food industry's other favorite decree—that there are no bad foods, just bad diets.

There *are* bad foods. And figuring out how to construct a healthier diet demands a basic and broad-based acceptance of which foods those are. That's why the proliferation of refurbished, purportedly healthy processed foods like my high-fiber biscotti is so vexingly problematic.

It's been said that if everyone were actually to consume the levels of fruits, vegetables, and whole grains recommended by the U.S. Dietary Guidelines, there wouldn't be enough to go around.

We'd have a serious shortage of broccoli. While this is true at the present moment, the various industries producing our food are not fixed. If everyone stopped eating so many french fries and started enjoying baked potatoes, stopped drinking orange beverages and started eating oranges, the system would reorder according to the basic principles of supply and demand. Blueberries, for instance, would become less expensive as more acres of them were planted (though they'd probably never be quite as cheap as corn and soy are today, due to their affinity for colder climates). The Pepsis, Kellogg'ses, Krafts, McDonald'ses, and Subways of the world might still be around, just with fundamentally different business models or operations on a much smaller scale. You need only look at what we ate a hundred years ago to see that in our modern world, food systems are always changing.

Who knows what the future holds? Who would have thought a hundred years ago that soybeans could become something like grilled chicken or that starches could transform into fat substitutes or biscotti-enhancing fiber? From medical imaging to antibiotics and iPhones, so many technological developments have served to enhance our lives. And while the marriage of technology and food has had enormously positive effects on food's availability and diversity, it hasn't been an unalloyed societal good. My hope is that through a greater understanding of the distorted journey our food undergoes, we'll arrive at a more enlightened and cautious understanding of the limits of food science, and an acceptance of the virtues of interacting with food according to the simpler methods so fundamentally intertwined with our fundamental biology. After all, most foods don't need to be "advanced through science." They're already pretty advanced the way they are.

Epilogue: A Note on the Paperback

I WROTE *PANDORA'S LUNCHBOX* AS an argument for eating foods that are closer to the ground. Most of us don't associate dirt with nourishment, but along with sunlight, it is the basis for all the foods that come from the earth and every molecule that offers us sustenance, even the ones we don't quite understand yet. Much of this nutrition delivered by the soil, or by animals that ultimately live by it, gets lost or discarded in the high-tech poking and prodding that happens when food is highly processed, diminishing food's essential nourishment and thus our health.

Since the book's publication, I've heard from many people who understand the intimate link between food and health, both for themselves and our entire overfed, malnourished nation. Others, though, are just learning this. My favorite interactions have been with people who until recently never paid much attention to what they ate and never questioned the insidious ubiquity of processed food; people for whom *Pandora's Lunchbox* was part of an eye-opening, life-changing journey. Thirty-three-year-old Jonathan Dembinski described himself as someone who has had "obesity issues" since childhood. He told me: "I have lost thirty-six pounds, and people keep asking me what diet I am on. I have to explain that I am not on a diet and that I have simply switched to real, nonprocessed foods. My body has responded positively to the change, and I have never felt this great in my life." Lydia from Sterling Heights, Michigan, wrote, "My eyes are opened to the world of processed food in a way that has completely changed the way I am eating and feeding my family." What strikes me about these accounts, and those of many others, is not only how much better Jonathan and Lydia say they feel from eating greater amounts of fresh, simple foods, but how quickly it can happen. Too often, eating well gets cast in the framework of long-term benefits, something you need to do to avoid dying too early. But chances are the payoffs are weeks, not years, away.

Eyes Wide Open

It's increasingly clear that something we might loosely call a "food movement" is under way in America. People are focusing more on buying local and planting gardens and avoiding meat raised with antibiotics and cutting back on processed food. The movement's unifying force is an unwillingness to eat blindly, to trust that mega food companies know what they're doing when it comes to feeding us.

It's obvious to me that this conscious eating movement is gaining momentum. Take the ingenuity of a (then) fifteen-year-old named Sarah Kavanagh, who in the fall of 2012 started a petition asking Pepsi to remove brominated vegetable oil (BVO) from Gatorade. BVO was the subject of negative safety studies in the 1970s, but the ingredient was allowed to stay on the market pending further study, which hasn't happened. Kavanagh's thought was:

Why would I want a strange and potentially dangerous food additive in my favorite drink? Lots of people—206,665 of them—had the same thought and signed Sarah's petition. Less than three months later, Pepsi announced it would remove BVO, which contains the toxic element bromine and can accumulate in fatty tissue, from Gatorade.

This story illustrated the power ordinary people can have, thanks to the Internet, to nudge big companies to change. Although on some level, the lesson is largely symbolic. Even without BVO, Gatorade is still glorified sugar water dressed up with artificial food dyes, which are other ingredients that shouldn't have free rein in our food supply.

Two months after Pepsi's announcement, Whole Foods said that by 2018 it would require all products sold in its U.S. and Canadian stores to indicate whether they contain genetically modified ingredients. Ever since GMO crops were first introduced into the food supply in 1996, the FDA and the mainstream food industry have been in lockstep agreement on the issue of GMO labeling: There's absolutely no need for it, they say, because genetically modified crops are no different than conventional ones. The public, which for quite some time didn't even realize that genetically modified corn, canola, and soy crops had found their way into most processed food, now begs to differ. Polls find that at least 80 percent of Americans believe that the presence of GMOs does in fact represent a distinction and should be identified on packages or labels. Even those who say they think GMOs are probably safe still think labels should exist and that shoppers should be able to decide for themselves whether the inclusion of GMO ingredients matters.

Whole Foods' decision marks the first acknowledgment (but certainly not the last) by a national retailer that a nearly two-decade-long lack of transparency has become unacceptable. My own view on GMOs is that adequate safety testing has yet to be done, largely because Monsanto and DuPont, the companies that make biotech seeds, haven't been willing to give independent researchers unfettered access to their products. As a result, it's impossible to determine whether these creations of advanced technology are acceptable for us and for the environment. I'm with the 80 percent. When you consider that GMOs—the ones currently on the market, not the hoped-for varieties with the soaring story lines about ending world hunger—offer zero benefit to the people eating them and minimal, if any, benefit to larger society, it's not hard to conclude that Americans should be allowed the choice of avoiding them. You can pick up food packages and get all manner of information about what's inside. GMOs need to be part of that.

It's encouraging that so many Americans are eager to know the story behind their food, since such curiosity tends to lead to the consumption of less junk and increased amounts of the good stuff, which, it should be noted, is slowly becoming more available. Every year there are greater numbers of farmers' markets in the U.S., and healthy food is turning up in unlikely places. I recently found myself marveling at what used to be a nutritional wasteland—the gas station convenience store. At several stores along the East Coast, I noticed there were items for sale that weren't Cheetos, Mountain Dew, hot dogs doing endless somersaults on greasy rollers, or boxes of those strangely small donuts. These stores had what appeared to be new sections featuring salads, wraps, baby carrots, fruit cups, small containers of hummus, and individually wrapped cheese. It was now possible to be on the road and stop in for a quick snack or lunch that might actually leave you feeling satiated, not urgently craving more sugary or salty stuff a little while later. I can only hope that sales of these foods will be such that more convenience stores decide to stock them.

There are other inspiring changes, too. At the government level, the USDA's new school lunch rules, although not exactly ushering in an era of fresh food and from-scratch cooking, will nonetheless likely increase the amount of fruits, vegetables, and whole grains that kids eat. In particular, the agency's proposed rules regarding vending machines in schools (which hopefully will be implemented) deserve applause because they will replace much of the crap that now tumbles out—including all regular sodas— with healthier choices like water, dried fruit, and nuts.

Still, for all the momentum those working in the food movement have gathered, there remain monumental levels of resistance, ignorance, and inertia. The year 2013 was, after all, one in which Carl's Jr. took Pop-Tarts and made them into ice cream sandwiches and Dunkin Donuts figured out what its bacon and egg sandwiches were missing all along—sugar. The chain swapped out the boring old bun for a glazed donut. And all this followed Taco Bell's marriage of Doritos with tacos in 2012. More than one hundred million Doritos Locos Tacos were sold in just ten weeks, the explosive demand for cheap tacos in a neon shell (thanks, Yellow No. 5 and Red No. 40 food colorings!) exceeding the chain's wildest expectations. Many Americans see no problem with this and simply aren't aware of the benefits of reading ingredient labels or considering how their food is made or grown.

Changing something as personal and deep-seated as eating habits is a slow, uphill climb. But reforming our food culture can be done. For evidence, look no further than how the food industry has so completely altered what we eat over the last half century.

The Path Ahead

The only real, lasting way to effect change in the food industry is for us, the customers, to vote with our food dollars. Companies must, out of necessity, listen to their customers. If we stop buying Toaster Strudel and fast-food sausage biscuits in favor of food without epic ingredient lists and dire health consequences, they will alter what they produce, probably dramatically. Many companies would likely merge to avoid going out of business, and the processed food industry would be smaller than it is today, perhaps more like what it is in Europe ($2,682 per person annually, versus $3,185 in the U.S.).

Our children represent the biggest reason for hope. Kids, at least up to a certain age, are naturally curious, and teaching them about fresh food and healthy ingredients can make a big difference in how they eat throughout their lives. So many of us consume a diet drowning in processed food because that's all we've ever known. Educating kids about the value of healthy foods before they develop a taste for soda, fast food, and Count Chocula is critical. Some schools now have gardens that teach kids about how vegetables grow. Cooking classes, be they after school, at summer camp, or conducted by a religious institution, are a great place for kids to learn how to have fun with food, ensuring they don't go through life not knowing what to do with an egg.

There are also great resources online to help teach kids. Two recent additions worth mentioning: *Mr. Zee's Apple Factory*, a free e-book that teaches young kids about how processed food is made and cleverly marketed to them, and Unjunk Yourself, a series of highly watchable music videos about the pointlessness of eating food that's "deep fried, modified, hydrogenized."

Along with education, key public policy changes would help

parents to feel less like imparting healthy eating habits to their kids is like swimming upstream in class 5 rapids. I don't let my kids, who are aged seven and four, watch any TV with commercials, since many of those ads are hawking sugary junk. But there will come a time when they're going to ignore my rules and be exposed to it anyway, at which point they will start asking me when I am going to buy them SpongeBob Mac & Cheese and Capri Sun Roarin' Waters. And with the food industry spending nearly $2 billion a year on marketing aimed directly at kids, that's a lot of SpongeBob, not to mention online Lunchables games and Oreo videos by country singer Kacey Musgraves. The packaged food industry states that it has nutritional standards in place so that only healthier food gets marketed to kids. But somehow those "nutritional standards" still allow for things like Kellogg's Chocolate Krave cereal, whose first ingredient is "Chocolate Flavored Filling." If more people started advocating for real, stringent restrictions on junk-food marketing to kids, whether through their elected representatives or Sarah Kavanagh–style via an online petition, the government agencies in a position to take action might actually summon the will to do so.

Other policy reform worth advocating for is a change to the USDA's food-stamp program, officially known as SNAP, so that its low-income recipients, who are disproportionately affected by obesity and diabetes, can't buy soda and other sugary beverages using taxpayer money. Let's face it: soda isn't food, not even a little bit. The whole reason food stamps were created in the first place was to ensure that economically disadvantaged Americans wouldn't starve or face malnutrition. Soda doesn't help with this endeavor. In an age of obesity and diabetes, the transfer of millions of dollars from taxpayers to soda companies makes no sense, something New York Mayor Michael Bloomberg recognized several years ago when he tried to get the USDA to let him prevent

New Yorkers from buying soda using food-stamp dollars. The USDA, a government agency never eager to butt heads with the food industry, said no. Instead, Bloomberg's administration has adopted other programs, like one that gives overweight and obese low-income residents money to be used at farmers' markets. It's a program administered through doctors and paired with nutrition counseling. If federal agencies won't budge, the future of public health might be in the hands of local entities.

The Snack Mindset

A few years ago Nestlé, the Swiss packaged food maker, loaded up a barge with its various candy bar, ice cream, cookie, boxed cereal, and packaged juice products and sent it down two Amazon River tributaries and into villages in Brazil's impoverished northeast regions. A floating supermarket, the barge docked in places where people had never had regular access to the pleasures of an ice cream Drumstick, for example. For Nestlé, which was looking to expand its business in Brazil and other developing countries, this was the only way to get packaged food to the people living in these areas, which lack supermarkets and grocery stores. It was a particularly swashbuckling way of doing what every major processed food company is now trying to do—reach people around the world, mostly those in developing countries that haven't been consumers of their food before.

Each year, the American and European (but mostly American) processed food industries export more and more of their products overseas to populations where traditional, home-cooked, mostly healthy meals still prevail. Because, while there is considerable ruckus here in the U.S. about the health effects of fast food and junk food, there is far less uproar in countries like Brazil,

India, China, and Russia, and the processed food industry has quietly set its sights on changing the way people in these highly populous countries eat. Public health in developing nations is arguably an even bigger concern than the state of the American diet.

This push for international expansion is nothing more than a financial necessity for food companies. As publicly traded corporations, they need continual growth in revenues and profits, and, with the American diet already consisting of 70 percent processed food, they're not going to get it from people in LA and Fort Lauderdale. New customers are to be found in Tianjin, Sao Paulo, and Bangalore, where Taco Bell opened up its first Indian location in 2010, much to the delight of the more than three thousand people who showed up on that first day. Yum Brands, the Kentucky-based parent company of Taco Bell, Pizza Hut, and KFC, already gets nearly three-quarters of its profits from international markets, many of those in the developing world. If you think there are too many KFC restaurants in the U.S., consider that there are now more in China. In the last two decades, Yum and its franchisees have opened 4,400 KFCs in 850 Chinese cities.

Nobody sat down fifty years ago and conspired to make Americans fat and beset with chronic diseases. Similarly, there is no one inside food companies who wants to see the Third World infected with the diseases of the First. But by flooding developing regions with green-tea-flavored Oreos, Tang, and the Colonel's Secret Recipe, that's what they're likely to do. The unintended consequences of global corporate growth are already evident and depressing. China, for instance, has just passed the U.S. for Type II diabetes prevalence: 11.6 percent of Chinese adults have the disease, compared with 11.3 percent here. In 1980, the diabetes rate in China was below 1 percent. When Asians gain weight, it appears that their risk for diabetes increases to a greater extent than it does for blacks or whites. Worldwide, obesity rates have doubled since 1980.

Global public health experts already have a name for these grim developments: nutrition transition. Of course, the phenomenon is bigger than simply processed food: it's also about increased meat consumption due to higher incomes and less home cooking as people move from rural areas to urban ones. But prefabbed, precooked, sugary, salty, and fatty food is no small part of the problem, because what the processed food companies, particularly the packaged food makers, are bringing to the developing world is not just new types of food, but a distinctly American way of eating— namely snacking. Kellogg's CEO John Bryant has called it "the snack mindset." His company and others like Kraft, General Mills, and Pepsico need to foster this type of thinking, because what they're selling will only get consumed if people start eating at all hours of the day and in between meals. Pringles, Pop-Tarts, Nutri-Grain bars, Cheese Nips: these are not proper meals.

Food companies respond to what little criticism and scrutiny they get over their international practices by saying that Pop-Tarts, Oreos, and buckets of fried chicken are all meant to be occasional treats that can fit perfectly well into a healthy diet. While this is true in theory, it's unfortunately not how it usually works. The millions of dollars food companies spend on slick marketing combined with the sugary, salty allure of processed food are powerful. The result is that hyperpalatable, portable, nutritionally vacuous foodstuffs replace real, nutritious meals. By necessity, the food industry's goals are not to sell their food occasionally, but to get as many people eating as much of it as possible. We've seen the health outcomes of this in the U.S., and if nothing is done, the developing world will be eating its way into the same mess.

The Food Industry as a (Small) Force for Good

Although the major food manufacturers have been largely silent about recent criticism of processed food, they are working furiously behind the scenes to make healthier products. My view is that this is not going to help very much. The conversations I've had with food scientists who contacted me after the book was published have reinforced this notion. They spoke about what they perceive to be a very limited ability to create healthier, fresher food, and they described their roadblocks as almost entirely economic. They are crunched, they said, between ingredient costs and the pressure their companies get from retailers, especially Walmart, the nation's largest food seller, to keep food prices extremely low. "Many, many products die due to cost," wrote JK, a longtime food scientist currently working for a natural and organic food maker. "Most innovative products never get out. I can literally put anything you can make in your kitchen into a package, with nearly everything in it near natural. But consistently, at least 90 percent of products are pushed to the wayside."

The companies caught most tightly in the vise of modern American economics are the packaged food makers such as Kraft, General Mills, Pepsi, and Kellogg's. The pressures they face from all fronts—shelf-life requirements, cost demands from retailers, low price expectations from shoppers, investors who want ever-increasing revenues and profits—explain why their "healthy" contributions to the national diet end up being things like Hot Pockets with whole grains and Lunchables with fruit smoothies. Most of the time, all these companies can do is create less terrible versions of their existing products.

Restaurant chains, however, offer more hope, at least in theory. Because of their ability to utilize a kitchen and thus serve reasonably fresh food, restaurants have a fighting chance at becoming part of a changed food landscape. It's worth giving modest applause, for instance, to McDonald's for the addition of oatmeal to its breakfast menu, and more recently for its McWraps, which contain actual vegetables (albeit in minor quantities) and chunks of relatively healthy chicken (assuming you don't get it "crispy," which means fried). The McWrap doesn't exactly go all the way to nutritious, since the tortilla is forty grams of blood-sugar-spiking white flour, plus highly processed ingredients like interesterified soybean oil and sodium metabisulfite. Carl's Jr., home of the Pop-Tart sandwich, now has a Charbroiled Cod Sandwich featuring a tasty, nonfried piece of fish. Upstart chains like Noodles & Company are selling freshly cooked pasta and vegetables. And of course there's Chipotle, which makes simple tacos and burritos on-site at its restaurants without additives (except for the tortillas and chips, which aren't made in the restaurants). Unfortunately, though, some of Chipotle's food manages to deliver half a day's recommended allowance of calories in one menu item, which isn't going to help anyone get healthy or manage his or her weight.

So yes, big companies can make a contribution. Maybe one day it will be possible for us to consume a steady supply of nutritious food that big, efficient companies cook for us. But for now, at least, the path to feeling good from day to day and to staying clear of dialysis machines, heart stents, and prescription pain pills runs through a place that's long been a center point of human living—our own kitchens.

Notes

B elow is a list of the sources I used to help write this book. It is by no means exhaustive. Instead, I wanted to provide an account of the primary sources I used and to document where much of the information in the book comes from. All references to ingredients originate from package labels, brand Web sites, or restaurant ingredient lists posted on company sites. Web site URLs are current as of November 2012.

Introduction

xiv *Americans are a different dietary species*: The turn-of-the-century portion of these statistics comes from Shirley Gerrior, Lisa Bente, and Hazel Hiza, *Nutrient Content of the U.S. Food Supply, 1909–2000*, Home Economics Research Report 56 (Washington, DC: U.S. Department of Agriculture, Center for Nutrition Policy and Promotion, 2004), a very useful report. Data on current consumption comes from "What We Eat in America, NHANES 2007–2008," U.S. Department of Health and Human Services and U.S. Department of Agriculture, July 30, 2010, available at http://www.ars.usda.gov/ba/bhnrc/fsrg; *2010 Dietary Guidelines for Americans* (Washington, DC: U.S. Department of Health and Human Services and U.S. Department of Agriculture, 2011), available at www.dietaryguide-lines.gov.

xv *Whole Foods co-founder John Mackey*: Katy McLaughlin and Timothy W. Martin, "As Sales Slip, Whole Foods Tries Health Push," *Wall Street Journal*, August 5, 2009.

xv *some 70 percent of our calories come from this*: This figure comes from Professor Carlos Monteiro of the Department of Nutrition at the School of Public Health, University of Sao Paulo, and his PhD student Larissa Baraldi, who were kind enough to send me a preliminary estimate of their work, which is based on 2007 and 2008 NHANES data. Although it's often mentioned that 90 percent of the money Americans spend on food goes to processed food, this figure includes things that shouldn't actually be classified as processed food. Monteiro's work is unique in that he divides food into three categories: fresh or minimally processed foods (including, for instance, frozen vegetables), processed culinary ingredients (such as sugar and other sweeteners, oils, and flours), and ready-to-consume ultraprocessed products. Ultraprocessed products are substances that have salt and/or sugar, a multitude of other ingredients including preservatives and cosmetic additives, and little or no whole food. These include breakfast cereals; energy bars; instant packaged soups and noodles; sweetened breads and buns, cakes, pastries, and desserts; chips and many other types of sweet, fatty, or salty snack products; and sugared fruit drinks, sodas, and energy drinks. Articles published by Monteiro and his colleagues in the journal *Public Health Nutrition* in 2011 and 2012 showed that processed and ultraprocessed products comprise only 20 percent of the calories consumed at home in Brazil, but 58.2 percent in the U.K. and 61.7 percent in Canada.

xv *As an industry, this amounts to $850 million a year*: This figure represents 70 percent of the total U.S. at-home and away-from-home food sales tracked by the USDA's Economic Research Service.

1. Weird Science

2 *prompted the novelist Tom Robbins to write*: Tom Robbins, *Jitterbug Perfume* (New York: Bantam Books, 1984).

6 *an ambitious campaign National Starch ran*: The "Starchology" ads ran in 2010 in various trade publications and Web sites, and are still viewable online at http://www.foodinnovation.com/starchexperts/.

6 *The amount we pay for our food has declined*: "Food Expenditures by Families and Individuals as a Share of Disposable Personal Income, Table 7," in the ERS Food Expenditure Series, Economic Research Service of the U.S.

Department of Agriculture, October 1, 2012, available at http://www
.ers.usda.gov/data-products/food-expenditures.aspx#26636;   Alyssa
Battistoni, "America Spends Less on Food Than Any Other Country,"
Mother Jones.com, February 1, 2012, available at www.motherjones
.com/blue-marble/2012/01/america-food-spending-less.

7 *catapulted to a quarter of all yogurt sales*: Alan Rappeport, "US Food Groups
Develop Taste for Greek Yoghurt," *Financial Times,* January 3, 2012.

7 *Greek yogurt must be strained in $10-million machines*: Conversation with
Paul Petersen, National Starch (Ingredion), at IFT 11, New Orleans.

13 *Purdue is one of only thirty-eight universities*: IFT maintains a list of accred-
ited undergraduate food science programs, available at http://www.ift
.org/knowledge-center/learn-about-food-science/become-a-food-scien-
tist/approved-undergrad-programs.aspx. Most universities that offer
undergraduate degrees also offer graduate ones.

14 *The first department began in 1918*: From an account of department history on UMass Amherst's food science Web site, available at http://www
.umass.edu/foodsci/clydesdale-center/history.html; and Walter Chenoweth, "The Birth of a New Science," *Phi Kappa Phi Journal* (December
1945).

17 *women now account for 65 percent of students*: These are my estimates based
on conversations with food science department heads and data from the
IFT Student Association.

18 *IFT didn't have its first female president until 1997*: "Celebrating IFT's 50th
Anniversary," *Food Technology* 43, no. 9 (September 1989).

2. The Crusading Chemist

To tell the story of Harvey Wiley I relied on a number of sources, including profiles and articles on FDA Web sites and Wiley's autobiography:
Harvey Wiley, *Harvey W. Wiley: An Autobiography* (Indianapolis: Bobbs-
Merrill, 1930). Among the other useful sources, see Harvey Wiley, *Foods
and Their Adulteration: Origin, Manufacture, and Composition of Food Products*
(Philadelphia: P. Blakiston's Son & Co., 1907); Oscar E. Anderson, *Health
of a Nation: Harvey Wiley and the Fight for Pure Food* (Chicago: published for
the University of Cincinnati by the University of Chicago Press, 1958);
Carol Lewis, "The 'Poison Squad' and the Advent of Food and Drug Regulation," *FDA Consumer* (November–December 2002); Wallace Janssen,
"The Squad that Ate Poison," *FDA Consumer* (December 1981–January
1982); William MacHarg, "Speaking of Dr. Wiley," *Good Housekeeping*

(April 1920); Harvey Wiley, *1001 Tests of Foods, Beverages, and Toilet Accessories, Good and Otherwise: Why They Are So* (New York: Hearst's International Library Co., 1914); Harvey Wiley, *Influence of Food Preservatives and Artificial Colors on Digestion and Health* (Washington, DC: U.S. Government Printing Office, 1904–1908); "Beef Made Troops Ill: Volunteers Testify before Army Court of Inquiry," *New York Times,* March 3, 1899.

30   *a Heinz advertising campaign characterized sodium benzoate*: Ann Vileisis, *Kitchen Literacy: How We Lost Knowledge of Where Food Comes from and Why We Need to Get It Back* (Washington, DC: Island Press, 2007).

31   *a lawsuit accusing Coca-Cola of misbranding*: Ludy T. Benjamin, "Pop Psychology: The Man Who Saved Coca-Cola," *Monitor on Psychology* (American Psychological Association) 40, no. 2 (February 2009).

32   *In one piece, he condemned new forms of ice cream*: Harvey Wiley, "Ice Cream and Iced Drinks," *Good Housekeeping,* July 1912.

33   *his estimation of white flour*: Harvey Wiley, "Dr. Wiley's Question-Box," *Good Housekeeping,* January 1918.

33   *the American Medical Association (AMA) held white bread*: Harvey Levenstein, *Paradox of Plenty: A Social History of Eating in America* (New York: Oxford University Press, 1993).

36   *saccharin has had a roller-coaster record of safety*: Jesse Hicks, "The Pursuit of Sweet: A History of Saccharine," *Chemical Heritage Magazine,* Spring 2010.

37   *implicated in childhood hyperactivity*: Donna McCann et al., "Food Additives and Hyperactive Behaviour in 3-Year-Old and 8/9-Year-Old Children in the Community: A Randomised, Double-Blinded, Placebo-Controlled Trial," *Lancet* 370, no. 9598 (2007): 1560–67.

3. The Quest for Eternal Cheese

Strangely, little has been published on the life of James Lewis Kraft and the early years of his company. I am grateful for the few sources available: Anne Bucker and Melanie Villines, *The Greatest Thing since Sliced Cheese: Stories of Kraft Food Inventors and Their Inventions* (Northfield, IL: Kraft Food Holdings, 2005); Arthur Baum, "A Man with a Horse and a Wagon," *Saturday Evening Post,* February 17, 1945; Mark William Wilde, "Industrialization of Food Processing in the U.S. 1860–1960," PhD dissertation, University of Delaware, 1988 (despite its unpublished status, this is an excellent account of just how new and disruptive Kraft's processed cheese was); Walter Price and Merlin Bush, "The Process Cheese

Industry in the United States: A Review," *Journal of Milk Food Technology* 37, no. 4 (1974); Joe Hermolin, "The Kraft Family and Kraftwood Gardens," *Wisconsin Magazine of History,* Summer 2010.

40   *As Kraft wrote in the patent he received in June 1916*: James Kraft, Process of Sterilizing Cheese and an Improved Product Produced by Such, U.S. Patent 1,186,524, filed March 25, 1916, and issued June 6, 1916.

43   *now 7 pounds per person per year on average*: "Per Capita Consumption of Selected Cheese Varieties, 1980–2010," Economic Research Service of the U.S. Department of Agriculture.

44   *Kraft noticed that milk protein concentrate*: The account of Kraft Singles and milk protein concentrate comes from conversations with cheese experts at the Wisconsin Center for Dairy Research and other sources in the cheese industry.

44   *the agency sent a letter to Betsy Holden*: Kraft Foods FDA Warning letter, December 18, 2002, available at http://www.fda.gov/ICECI/Enforcement Actions/WarningLetters/2002/ucm145363.htm.

45   *with just 3 percent of its weight coming from additives*: Conversation with Lloyd Metzger, professor of dairy science, South Dakota State University.

48   *bioactive peptides, substances thought to ward off infection*: L. Seppo et al., "A Fermented Milk High in Bioactive Peptides Has a Blood Pressure–Lowering Effect in Hypertensive Subjects," *American Journal of Clinical Nutrition* 77, no. 2 (2003): 326–30.

50   *"Food must not be dead. It must have a soul"*: Harvey Wiley, "The Campaign for Pure Food," *New York Medical Journal,* August 4, 1917.

4. Extruded and Gun Puffed

51   *Then we prepare our cereals in a method similar to what you do*: "From Seed to Spoon," Kellogg's Web site, available at http://www.kelloggs.com /en_US/the-goodness-of-grains/from-seed-to-spoon.html.   For the story of the Kellogg brothers and the birth of modern breakfast cereal, I relied upon several excellent accounts: Scott Bruce and Bill Crawford, *Cerealizing America: The Unsweetened Story of American Breakfast Cereal* (Boston: Faber and Faber, 1995); *Corn Flake Kings,* A&E Home Video (2005); Marty Gitlin and Topher Ellis, *The Great American Cereal Book: How Breakfast Got Its Crunch* (New York: Abrams Image, 2011).

53   *on any given day that isn't Cereal Fest, one fifth of them*: Research from the Dairy Research Institute shows that ready-to-eat cereal with milk is a predominant breakfast pattern for one third of children and 20 percent

of adults. This represents average daily consumption. The institute did not look at how frequently within a typical week cereal with milk is consumed at breakfast.

53 *$10 billion annual U.S. business*: "U.S. Breakfast Cereals," Mintel, February 2012.

53 *Items from the cereal aisle are the eighth most popular*: "State of the Industry Almanac," *Grocery Headquarters*, April 2012, 40.

54 *more than half the world's breakfast cereal consumption*: E. J. Schultz, "Cereal Marketers Race for Global Bowl Domination," *Advertising Age*, August 20, 2012.

58 *"Made with wholesome grains," says Kellogg's*: "The Benefits of Cereal," Kellogg's Web site, available at http://www.kelloggs.com/en_US/the-benefits-of-cereal.html.

59 *In 1905, he changed the Corn Flakes recipe*: E-mail correspondence with Kellogg's; brief historical account on the company's Web site.

59 *The first transformative technology*: Bruce and Crawford, *Cerealizing America*.

60 *"The Miracle at Your Table"*: Mark William Wilde, "Industrialization of Food Processing in the U.S. 1860–1960," PhD dissertation, University of Delaware, 1988.

60 *Grape-Nuts bore this invitation*: From images of 1905 boxes printed in Gitlin and Ellis, *Great American Cereal Book*.

61 *a map shows the exact location*: Bear River Valley Web site, available at http://www.bearrivervalleycereal.com/bear-river-valley/.

61 *They're made eight miles north in Tremonton*: Conversations with Linda Fischer, MOM Brands consumer marketing manager, and company customer service reps.

61 *After gun puffers came extrusion machines*: Extrusion manufacturers were helpful in explaining how their machines work and how they're used. I talked with Brian Hinkle, market manager at Buhler; Brian Plattner, process engineering manager at Wenger; and Bill Butler, sales manager, and Jean-Marie Bouvier, vice president of technology, both at Clextral. See also Mian Riaz, ed., *Extruders in Food Applications* (Lancaster, PA: Technomic Publishing, 2000); Lynn A. Kuntz, "Breakfast Cereals Grow Up," *Food Product Design*, August 2000; Tom Dworetzky, "The Churn of the Screw," *Discover*, May 1988.

63 *According to a 2009 study done by Mian Riaz*: Mian Riaz et al., "Stability of Vitamins during Extrusion," *Critical Reviews in Food Science and Nutrition* 49, no. 4 (2009): 361–68.

64 *So, too, with naturally occurring flavor*: Riaz, *Extruders in Food Applications*.

65 *8 percent of the U.S. population suffers from type II diabetes*: "Number of Americans with Diabetes Rises to Nearly 26 Million," Centers for Disease Control press release, January 26, 2011.

66 *between 5 and 15 percent of a person's daily calorie expenditure*: Klaas R. Westerterp, "Diet Induced Thermogenesis," *Nutrition & Metabolism* 18, no. 1 (2004): 5.

66 *A team of Japanese researchers fed two groups of young rats*: Kuniyuki Oka et al., "Food Texture Differences Affect Energy Metabolism in Rats," *Journal of Dental Research* 82, no. 6 (2003): 491–94.

66 *A California study that evaluated human responses*: Sadie B. Barr and Jonathan C. Wright, "Post-Prandial Energy Expenditure in Whole-Food and Processed-Food Meals: Implications for Daily Energy Expenditure," *Food & Nutrition Research* 54 (July 2010): 5144. DOI: 10.3402/fnr .v54i0.5144.

67 *Which is exactly what Robert Choate did*: Harvey Levenstein, *Paradox of Plenty: A Social History of Eating in Modern America* (New York: Oxford University Press, 1993); Bruce and Crawford, *Cerealizing America*; Elaine Woo, "Robert B. Choate Dies at 84; Consumer Advocate Pushed for Healthier Cereals," *Los Angeles Times*, May 17, 2009; Jack Rosenthal, "Hunger Expert Says Many Dry Cereals Are Not Nutritious," *New York Times*, July 24, 1970; "Consumerism: Breakfast of Chumps?" *Time*, August 3, 1970; "Accusations Denied by Cereal Firms," *Toledo Blade*, July 24, 1970; Richard Halloran, "Breakfast Cereal Industry Tells Senators Nutrition Critic Erred: Breakfast Food Industry Rebuts Critic," *New York Times*, August 4, 1970.

69 *a nutrition ratings program called Smart Choices*: William Neuman, "For Your Health, Froot Loops," *New York Times*, September 4, 2009.

70 *But Choate wasn't finished*: Transcript of Choate's testimony: "Too Stuffed for Supper," March 2, 1972, before Senate Commerce Committee, Subcommittee on the Consumer.

71 Consumer Reports *sought to replicate Choate's tests*: "Which Cereals Are Most Nutritious?" *Consumer Reports*, February 1975.

5. Putting Humpty Dumpty Back Together Again

74 *Every day at the kinetic Port of Melbourne, Australia*: E-mail correspondence with Donald Forsdyke, business development manager, Port of Melbourne.

75   *Firms there buy between 70 percent and 80 percent*: "Wool Grease Production: Going East," Lanolin.com/Imperial-Oel-Import (IOI) news brief, March 1, 2010.

75   *the grease's cholesterol component is used to make vitamin D*: Conversation at SupplySide West 2011 conference with Daniel Fang, vice manager, Zhejiang Garden Biochemical.

76   *Eijkman made the first critical discovery*: "Christiaan Eijkman, Beriberi and Vitamin B1," Nobel Prize.org, available at http://www.nobelprize.org /educational/medicine/vitamin_b1/eijkman.html.

76   *decided to pick up where the Dutchman had left off*: Paul Griminger, "Casimir Funk: A Biographical Sketch (1884–1967)," *Journal of Nutrition* 102 (1972): 1105–1114; "Casimir Funk and a Century of Vitamins," The Pauling Blog, June 15, 2011, available at http://paulingblog.wordpress .com/2011/06/15/casimir-funk-and-a-century-of-vitamins/.

77   *Kellogg's Pep became one of the first breakfast cereals*: "A Historical Overview," Kellogg's History Web site, available at http://www.kellogghistory.com /history.html/.

77   *ran an eye-catching ad campaign*: Letters between J. I. Sugerman of the Doughnut Corporation of America and Wilburn L. Wilson, assistant director in charge of nutrition at the Department of Defense, National Archives, Records of the Agricultural Marketing Service, 1942.

79   *It wasn't until 2005 that the U.S. Department of Agriculture's*: This is my estimation based on a review of Dietary Guidelines and Dietary Goals for the United States from 1977 to 2010.

79   *Our consumption of whole grains is up 20 percent*: "Whole Grain Statistics," Whole Grains Council Web site, available at http://www.wholegrain-scouncil.org/newsroom/whole-grain-statistics.

79   *13 percent of the total grains we eat*: Extrapolated from Hodan Farah Wells and Jean C. Buzby, "Dietary Assessment of Major Trends in U.S. Food Consumption, 1970–2005," Economic Research Service of the U.S. Department of Agriculture, March 2008. Americans on a 2,000-calorie-per-day diet consumed 8.1 ounce equivalents of grains per person per day in 2005, of which 7.2 ounce equivalents were refined grains and 0.9 ounce equivalents were whole grains.

79   *"Marshmallow Pebbles is a wholesome, sweetened rice cereal"*: Post Foods Web site, available at http://www.postfoods.com/cereals/pebbles/?id=marsh mallow#nutrition.

80   *the only synthetic ingredients with carte blanche approval*: Conversation with Gwendolyn Wyard, associate director of organic standards and industry outreach, Organic Trade Association.

80 *vitamins a $3 billion worldwide business*: "Vitamins: A Global Strategic Business Report," Global Industry Analysts, February 2011.

81 *vitamin C from acerola cherries*: Conversation with Antoine Dauby, Naturex group marketing director, at SupplySide West, 2011.

81 *Chinese firms manufacture 60 percent*: Elaine Watson, "China Dominates Additive Supply," Food Manufacture.co.uk, November 30, 2009, available at http://www.foodmanufacture.co.uk/Business-News/China-dominates-additives-supply.

82 *The last vitamin C plant operating in the United States*: Sara Leitch, "Layoffs Begin at White Twp. Plant," *Easton (PA) Express-Times,* October 4, 2005; "Roche Vitamins to Pay More than $200,000 for Air Violations," Environmental Protection Agency press release, July 10, 2002; Bryan W. Waagner, "N.J. Fines Roche Vitamins for Air Emissions Violations," *Easton (PA) Express-Times,* December 30, 2001.

82 *It's releasing similar toxins to Roche's Belvidere plant*: Toxic Release Inventory, Environmental Protection Agency, 2010, available at http://oaspub.epa.gov/enviro/multisys2.get_list_tri?tri_fac_id=77541HFFMN1000C.

83 *Vitamin C production is a convoluted operation*: The descriptions of how vitamins are produced come from conversations with a variety of people who work in the vitamin industry, including Michael DeGennaro, global head of sales and marketing at Lonza. Many of the others preferred not to have their names used. Additionally, there were a few useful printed sources: Volker Spitzer, ed., *Vitamin Basics: The Facts about Vitamins in Nutrition,* 3rd edition (Waldkirch, Germany: DSM Nutritional Products AG, 2007), available at http://www.dsm.com/en_US/downloads/dnp/Vitamin_Basics.pdf; "GMO-Free Additives and Processing Aids for Organic Food and Feed Production: An Exploratory Study of Obstacles and Solutions," LIS Consult and Bioconnect, December 2008; R. Chuck, "Green Sustainable Chemistry in the Production of Nicotinates," Lonza, Inc., September 2006.

85 *supplemental vitamins fail to offer the health benefits*: Neena Samuel, "The Vitamin Myth," *Reader's Digest,* November 2007; Matthew Herper and Rebecca Ruiz, "Snake Oil in Your Snacks," *Forbes,* June 7, 2010; Sarah Mahoney, "Time to Kick the Multivitamin Habit, Studies Suggest," *Prevention,* November 2010; S. M. Lippman et al., "Effect of Selenium and Vitamin E on Risk of Prostate Cancer and Other Cancers: The Selenium and Vitamin E Cancer Prevention Trial (SELECT)," *Journal of the American Medical Association* 301, no. 1 (2009): 39–51.

86 *One study in 2012, however, bucked the trend*: J. Michael Gaziano et al., "Multivitamins in the Prevention of Cancer in Men: The Physicians' Health Study II Randomized Controlled Trial," *Journal of the American Medical Association* 308, no. 18 (2012): 1871–80.

86 *a 2011 study in the* Journal of Nutrition: Victor L. Fulgoni III et al., "Foods, Fortificants, and Supplements: Where Do Americans Get Their Nutrients?" *Journal of Nutrition* 141 (2011): 1847–54.

88 *half of us are suffering from one or more chronic diseases*: Gerard Anderson and Jane Horvath, "The Growing Burden of Chronic Disease in America," *Public Health Reports* 119, no. 3 (2004): 263–70.

88 *Americans rank dismally for life expectancy at number 37*: Sandeep C. Kulkarni et al., "Falling Behind: Life Expectancy in U.S. Counties from 2000 to 2007 in an International Context," *Population Health Metrics* 9, no. 16 (2011), available at http://www.pophealthmetrics.com/content/9/1/16.

88 *Americans have also lost the first-place trophy for tallest population*: Rob Stein, "America Loses Its Stature as Tallest Country," *Washington Post,* August 13, 2007.

89 *the antioxidant activity of apples*: Marian V. Eberhardt et al., "Nutrition: Antioxidant Activity of Fresh Apples," *Nature* 405 (2000): 903–4.

90 *the specific components in wheat bran*: Y. Zhu et al., "5-alk(en)ylresorcinols as the major active components in wheat bran inhibit human colon cancer cell growth," *Bioorganic and Medicinal Chemistry* 19, no. 13 (2011): 3973–82.

91 *Sang grew up in Shandong, China*: The profile of Sang comes from my conversation and e-mail correspondence with him.

92 *eighty-nine-year-old billionaire named David Murdock*: Frank Bruni, "The Billionaire Who Is Planning His 125th Birthday," *New York Times Magazine,* March 3, 2011.

92 *Coke funded a study*: William Mullen et al., "A Pilot Study on the Effect of Short-Term Consumption of a Polyphenol Rich Drink on Biomarkers of Coronary Artery Disease Defined by Urinary Proteomics," *Journal of Agriculture and Food Chemistry* 59, no. 24 (2011): 12850–57.

94 *two studies she did involving broccoli*: J. D. Clark et al., "Bioavailability and Inter-Conversion of Sulforaphane and Erucin in Human Subjects Consuming Broccoli Sprouts or Broccoli Supplement in a Cross-Over Study Design," *Pharmacological Research* 64, no. 5 (2011): 456–63; J. D. Clark et al., "Comparison of Isothiocyanate Metabolite Levels and Histone Deacetylase Activity in Human Subjects Consuming Broccoli Sprouts

or Broccoli Supplement," *Journal of Agricultural and Food Chemistry* 59, no. 20 (2011): 10955–63.

94   *powder from whole tomatoes versus lycopene*: J. K. Campbell et al., "Tomato Phytochemicals and Prostate Cancer Risk," supplement, *Journal of Nutrition* 134, no. 12 (2004): 3486S–92S.

95   *The blueberry compound pterostilbene*: E-mail correspondence with Jerry Bartos, senior product development manager, Chromadex.

6. Better Living through Chemistry

101 *Subway stores get it frozen*: Conversation with Mark Christiano, Subway's global baking specialist.

101 *Tim Zagat told the* Today Show's *Matt Lauer*: Today Show, September 6, 2011.

101 *When subjected to industrial production*: My description of industrial baking comes from several sources in the industry, including Tom Lehmann, director of baking assistance, American Institute of Baking; Jan Van Eijk, research director, Lallemand, a baking ingredient supplier; Doug Radi, senior vice president of marketing and sales, and Rich Gallegos, plant manager, Rudi's Organic; and Scott Creevy at Great Harvest in Boulder.

102 *To create sodium stearoyl lactylate*: Addy Cameron Huff, "Sodium Stearoyl Lactylate: Modern Baking's Mystery Ingredient," University of Waterloo, March 2009, available at http://www.docstoc.com/docs/33668507/Sodium-Stearoyl-Lactylate.

103 *On a summer morning in 2001*: Gary Wisby, "Spill Halts CTA, Dan Ryan 17 Firefighters, Cops Are Hospitalized from Heat Exhaustion," *Chicago Sun-Times*, August 8, 2001.

104 *The agency often seeks to convey the impression*: "Food Ingredients and Colors," International Food Information Council and U.S. Food and Drug Administration, November 2004; reviewed April 2010, available at http://www.fda.gov/food/foodingredientspackaging/ucm094211.htm.

105 *An influential book published in 1933*: Arthur Kallet and F. J. Schlink, *100,000,000 Guinea Pigs* (New York: Vanguard Press, 1933).

105 *Representative James Delaney from New York*: "Food Additives Arouse Dispute," *New York Times*, June 30, 1957; "U.S. Ends Testing of 188 Food Items," *New York Times*, November 25, 1958; "Food Makers Given Until Feb. 1 to Ask Delay on Additive Law," *New York Times*, December 31, 1960.

105 *some two thousand additives allowed in food*: Peter Schuyten, "Technology: Finding Better Food Additives," *New York Times,* February 7, 1980.

105 *a first-rate report on food additives in 2011*: Thomas G. Neltner et al., "Navigating the Food Additive Regulatory Program," *Comprehensive Reviews in Food Science and Food Safety* 10 (2011); an invaluable overview and analysis of the confusing jumble that is food regulation.

106 *the illuminating entry for azodicarbonamide*: "Food Additive Status List," U.S. Food and Drug Administration Web site, available at http://www .fda.gov/Food/FoodIngredientsPackaging/FoodAdditives/FoodAdditiveListings/ucm091048.htm#ftnA.

106 *Congress expected that all new substances*: "History of the GRAS List and SCOGS Reviews," U.S. Food and Drug Administration Web site, available at http://www.fda.gov/Food/FoodIngredientsPackaging/GenerallyRecognizedasSafeGRAS/GRASSubstancesSCOGSDatabase/ucm 084142.htm; Paulette M. Gaynor et al., "Approach to the GRAS Provision: A History of Processes: Excerpted from Poster Presention at the FDA Science Forum—April 2006," Division of Biotechnology and GRAS Notice Review, Office of Food Additive Safety, Center for Food Safety and Applied Nutrition, Food and Drug Administration, available at http://www.fda.gov/Food/FoodIngredientsPackaging/GenerallyRecognizedasSafeGRAS/ucm094040.htm.

107 *The FDA said that instead of GRAS* petitions: "Substances Generally Recognized as Safe; Proposed Rule," *Federal Register* 62, no. 74 (April 17, 1997): 18938–64.

107 *formal FDA "food additive petitions"*: Conversation with Tom Neltner, Pew Charitable Trusts.

109 *customers called Kellogg's toll-free phone number*: "Kellogg Company Voluntarily Recalls Select Packages of Kellogg's® Corn Pops®, Kellogg's® Honey Smacks®, Kellogg's® Froot Loops® and Kellogg's® Apple Jacks®," Kellogg Company press release, June 25, 2010.

109 *elevated levels of an unstudied chemical*: Sonya Lunder et al., "Kellogg's Cereal Recall: Health Risks from Packaging?" Environmental Working Group, available at http://www.ewg.org/health-risks-from-packaging.

109 *Froot Loops could include as many as twenty other food-contact substances*: In the FDA's "Food and Nutrition" webinar on March 6, 2012, Martin Hoagland, an FDA toxicologist, presented this as a possibility (slide 162).

110 *less than half of all food additives*: Conversation with Tom Neltner, Pew Charitable Trusts, based on his analysis of CDC and FDA databases.

110 *Of the 415 substances studied, twenty-five were recommended*: "Panel Clears

Most Food Additives But Demurs on Salt and Caffeine," *New York Times,* December 31, 1980.

111 *has placed it on its list*: "Reasonably Anticipated to Be Human Carcinogens," in *Report on Carcinogens, Twelfth Edition,* U.S. Department of Health and Human Services, Public Health Service, National Toxicology Program (Washington, DC: Research Triangle Park, NC, 2010), available at http://ntp.niehs.nih.gov/ntp/roc/twelfth/ListedSubstancesReasonablyAnticipated.pdf.

111 *Over the years, various studies have raised concerns*: "Potassium Bromate," in *Chemical Safety Information from Intergovernmental Organizations,* International Programme on Chemical Safety 73 (1999): 481, available at http://www.inchem.org/documents/iarc/vol73/73-17.html.

111 *tests done in Britain in the late eighties*: Paul Kamman, "Baking Bread and Rolls without Potassium Bromate," *Bakery Production and Marketing,* March 24, 1991.

111 *the agency asked bakers to stop using potassium bromate*: Lisa R. Van Wagner, "1992 Economic Outlook for the Food Industry," *Food Processing,* February 1, 1992.

111 *a known carcinogen called semicarbazide*: Adam Becalski, "Semicarbazide Formation in Azodicarbonamide-Treated Flour: A Model Study," *Journal of Agricultural and Food Chemistry* 52, no. 18 (2004): 5730–34; Gregory O. Noonan et al., "The Determination of Semicarbazide (N-aminourea) in Commercial Bread Products by Liquid Chromatography-Mass Spectrometry," *Journal of Agricultural and Food Chemistry* 53, no. 12 (2005): 4680–85.

112 *FDA suggested that companies please just use a bit less of it*: Erin Rigik, "Enzymes 101," *Baking Management,* April 1, 2009.

112 *In animal studies, the chemical has been linked*: Toxnet Toxicology Data, National Institutes of Health, available at http://toxnet.nlm.nih.gov/cgi-bin/sis/search/a?dbs+hsdb:@term+@DOCNO+838.

113 *small amounts of benzene*: "Consumer Reports Finds Benzene in Some Beverages; FDA Should Restrict Benzene in All Beverages; Consumers Should Take Precautions," *Consumer Reports* press release, August 25, 2006.

113 *too much phosphate can place a heavy burden*: Eberhard Ritz et al., "Phosphate Additives in Food—a Health Risk," *Deutsches Ärzteblatt International* 109, no. 4 (2012): 49–55.

116 *hyperactivity and ADHD in kids*: Donna McCann et al., "Food Additives and Hyperactive Behaviour in 3-Year-Old and 8/9-Year-Old Children in

the Community: A Randomised, Double-Blinded, Placebo-Controlled Trial," *Lancet* 370, no. 9598 (2007): 1560–67.

116 *The FDA has looked at this conflicting data*: "FDA Advisors Want More Study of Food Dye-ADHD Link," The Chart blog, CNN, March 31, 2011, available at http://thechart.blogs.cnn.com/2011/03/31/more-research-needed-on -food-dyes-fda-panel-says/.

117 *Whole Foods list of seventy-eight "unacceptable food ingredients"*: "Unacceptable Ingredients for Food," Whole Foods Web site, available at http:// www.wholefoodsmarket.com/about-our-products/quality-standards /unacceptable-ingredients-food.

118 *Novozymes, a Danish company*: All information on enzymes and Novozymes comes from company Web sites and brochures, as well as conversations with company employees.

122 *Cornucopia found at least seven "natural" breakfast cereals*: "Cereal Crimes: How 'Natural' Claims Deceive Consumers and Undermine the Organic Label—A Look Down the Cereal and Granola Aisle," Cornucopia Institute, October 2011.

7. The Joy of Soy

125 *some 10 percent of daily calories*: This is my calculation based on data in the following paper: Loren Cordain et al., "Origins and Evolution of the Western Diet: Health Implications for the 21st Century," *American Journal of Clinical Nutrition* 81, no. 2 (2005): 341–54, Table 1: "Refined Vegetable Oils Constitute 17.6 Percent of Calories in U.S. Diet" (thus, with 62 percent of all vegetable oils being soybean oil, this fat accounts for roughly 10 percent).

125 *Some 55 percent of the entire state of Illinois*: Agricultural Statistics Service, U.S. Department of Agriculture, 2011: corn 12.5 million acres, soybeans 8.9 million acres; Illinois land mass is 58,000 square miles, according to the Illinois Department of Natural Resources.

126 *soybeans haven't been here for very long*: Kenneth F. Kiple and Kriemhild Conee Ornelas, eds., *The Cambridge World History of Food* (New York: Cambridge University Press, 2000); Soyinfo Center Web site, available at http://www.soyinfocenter.com/. Run by Willam Shurtleff and Akiko Aoyagi, this is an incredibly detailed account of soybean history.

126 *74 million acres of soybeans*: "Soybean Area Planted and Harvested—States and United States: 2010 and 2011," National Agricultural Statistics Service, U.S. Department of Agriculture.

126 *14 billion pounds of soybean oil*: E-mail correspondence with United Soybean Board.

126 *government farm subsidies totaling $1.5 billion a year*: "2012 Farm Subsidy Database," Environmental Working Group, available at http://farm.ewg.org/index.php.

127 *62 percent of the market for added oils*: E-mail correspondence with United Soybean Board.

127 *humans didn't consume vegetable oils in large quantities*: Kiple and Ornelas, *Cambridge World History of Food*.

127 *The only isolated vegetable oil eaten*: Cordain, "Origins and Evolution."

127 *82 percent of all of our fat came from animals*: Shirley Gerrior, Lisa Bente, and Hazel Hiza, *Nutrient Content of the U.S. Food Supply, 1909–2000*, Home Economics Research Report 56 (Washington, DC: U.S. Department of Agriculture, Center for Nutrition Policy and Promotion, 2004).

127 *soybean oil had dislodged butter*: Extrapolation from "Fats and Oils (Added)," in the Food Availability (Per Capita) Data System," Economic Research Service, U.S. Department of Agriculture, August 2012, available at http://www.ers.usda.gov/data-products/food-availability-(per-capita)-data-system.aspx.

128 *first went into mass production here in 1922*: Soyinfo Center Web site at http://www.soyinfocenter.com; A. E. Staley and ADM files at the Decatur Public Library.

130 *What happens inside ADM's Decatur factory*: The description of soybean oil extraction and production comes from a number of sources—conversations with people in the industry, the Ag Processing video, and the following published sources: "Food and Agricultural Industries, 9.11.1 Vegetable Oil Processing," in *Emissions Factors and AP 42: Compilation of Air Pollution Emission Factors*, volume 1, 5th edition (Research Triangle Park, NC: Office of Air Quality Planning and Standards, Environmental Protection Agency, January 1995), available at http://www.epa.gov/ttnchie1/ap42/ch09/final/c9s11–1.pdf; Allan Smith and Sidney Circle, *Soybeans: Chemistry and Technology*, vol. 1 (Roslyn, NY: AVI Publishing Co., 1978).

130 *Hexane is classified*: "Occupational Safety and Health Guideline for n-Hexane," Occupational Safety and Health Administration, U.S. Department of Labor, available at http://www.osha.gov/SLTC/healthguidelines/n-hexane/recognition.html.

130 *Chinese workers at a factory making iPhones*: David Barboza, "Workers Sickened at Apple Supplier in China," *New York Times*, February 22, 2011.

131 *has the potential to cause permanent nerve damage*: E-mail correspondence with Environmental Protection Agency.

131 *The vegetable oil industry is the largest emitter of hexane*: Toxic Release Inventory, Environmental Protection Agency.

131 *Ag Processing soy biodiesel facility*: "Three People Transferred to Sioux City Burn Unit after Port Neal Explosion," Associated Press, August 31, 2003; "Investigation into Sioux City Plant Explosion Inconclusive," Associated Press, December 6, 2003.

133 *ADM sells twelve different vitamin E products*: Food Ingredients Catalog, ADM, 2011–2012.

133 *A soap manufacturer received a special award*: "Ecolab Recognized with IFT Food Expo Innovation Award," Ecolab press release, June 13, 2011.

136 *published in the* Journal of the American Oil Chemists' Society: C. M. Seppanen and A. Saari Csallany, "Formation of 4-Hydroxynonenal, a Toxic Aldehyde, in Soybean Oil at Frying Temperature," *Journal of the American Oil Chemists' Society* 79, no. 10 (2002): 1033.

136 *When the study was published in the same journal in 2004*: C. M. Seppanen and A. Saari Csallany, "Incorporation of the Toxic Aldehyde 4-Hydroxy-2-Trans-Nonenal into Food Fried in Thermally Oxidized Soybean Oil," *Journal of the American Oil Chemists' Society* 81, no. 12 (2004): 1137.

137 *HNE concentrations ranging from 7 to 32 parts*: Correspondence with A. Saari Csallany. As of November 2012, the study had yet to be published.

138 *Several years ago, the company had researchers*: Sean LaFond et al., "Formation of 4-Hydroxy-2-(E)-Nonenal in a Corn–Soy Oil Blend: A Controlled Heating Study Using a French Fried Potato Model," *Journal of the American Oil Chemists' Society* 88, no. 6 (2011): 763–72.

138 *a trail of messy environmental PR*: Martin Hickman, "The Guilty Secrets of Palm Oil: Are You Unwittingly Contributing to the Devastation of the Rain Forests?" *The Independent*, May 2, 2009.

139 *K. C. Hayes, a biologist at Brandeis University, has done studies*: K. Sundram et al., "Stearic Acid-Rich Interesterified Fat and Trans-Rich Fat Raise the LDL/HDL Ratio and Plasma Glucose Relative to Palm Olein in Humans," *Nutrition & Metabolism* 15, no. 4 (2007): 3.

139 *Monsanto and DuPont are taking the fight upstream*: See DuPont's Web site for Plenish soybeans at http://www.plenish.com; and Monsanto's Web site for Vistive Gold soybeans at http://www.vistivegold.com.

140 *the government's Dietary Guidelines stated*: Report of the Dietary Guidelines Advisory Committee on the Dietary Guidelines for Americans, 2010, part D,

section 3, p. D3–D25, June 14, 2010, available at http://www.cnpp.usda.gov/dgas2010-dgacreport.htm.

140 *little to no correlation with heart disease*: Dariush Mozaffarian and David Ludwig, "Dietary Guidelines in the 21st Century—A Time for Food," *Journal of the American Medical Association* 304, no. 6 (2010): 681–82.

141 *one and three omega 6 fats to every one omega 3*: P. M. Kris-Etherton et al., "Polyunsaturated Fatty Acids in the Food Chain in the US," *American Journal of Clinical Nutrition* 71, no. 1 (2000): 179S—88S; Susan Allport, *Queen of Fats: Why Omega-3s Were Removed from the Western Diet and What We Can Do to Replace Them* (Berkeley, CA: University of California Press, 2006); an engaging and immensely thorough look at the evolution of the science of omega 3 and 6 fats.

141 *found linoleic acid levels*: Tanya L. Blasbalg et al., "Changes in Consumption of Omega-3 and Omega-6 Fatty Acids in the United States during the 20th Century," *American Journal of Clinical Nutrition* 93, no. 5 (2011): 950–62.

144 *a video called "What Is Food Science, Anyway?"*: College and Career Resources, Discovery Education Web site, available at http://school.discoveryeducation.com/foodscience/college_resources.html.

8. Extended Meat

145 *corporate executives and factory workers gathered*: Willam Shurtleff and Akiko Aoyagi, "Central Soya Company (1934–): Work with Soy," Soyinfo Center, available at http://www.soyinfocenter.com/HSS/central_soya.php; a great resource for the history of soy protein.

146 *Initially soy meal went primarily into industrial products*: "Navy Bean Soup," *Time*, December 6, 1943.

146 *Henry Ford famously made car parts*: Willam Shurtleff and Akiko Aoyagi, "Henry Ford and His Employees: Work with Soy," Soyinfo Center, available at http://www.soyinfocenter.com/HSS/henry_ford_and_employees.php.

147 *a series of puddings and nondairy desserts*: James J. Nagle, "Iowa Plant to Produce Soy Protein," *New York Times*, May 18, 1969.

147 *In 1969, a* New York Times *article explained*: Nagle, "Iowa Plant to Produce Soy Protein."

148 *Last year it sold roughly $1.3 billion*: E-mail correspondence with Jennifer Starkey at Solae.

150 *In a chart on its Web site*: "Customer Solutions: Maximizing Profitability," Solae Web site at http://www.solae.com/Soy-Solutions/Maximizing-Profit.aspx.

150 *Several years ago, the company helped*: "Customer Solutions: Delivering Results: Case Studies," Solae Web site at http://www.solae.com/Soy-Solutions/Case-Studies/Meat-Processing.aspx.

151 *In 2009, they lost $360,000*: Amy Bounds, "Boulder Valley Losing Money on Healthy School-Food Program," *Boulder (CO) Daily Camera*, September 13, 2010.

151 *On its K-12 Web site, the company boasts*: "K–12 Working at the Heart of Your School Menu: Unique Chicken Process," Tyson Food Service Web site, http://www.tysonfoodservice.com/K-12/Commodities/Unique-Chicken-Process.aspx. This quotation was accessed in 2011, and the passage has since been slightly rewritten.

152 *A USDA rule used to limit "vegetable protein product"*: "Modification of the 'Vegetable Protein Products' Requirements for the National School Lunch Program, School Breakfast Program, Summer Food Service Program and Child and Adult Care Food Program," *Federal Register* 65, no. 47 (March 9, 2000).

155 *Raw soybeans have thirty-three different flavor compounds*: Allan K. Smith and Sidney J. Circle, eds., *Soybeans: Chemistry and Technology,* volume 1 (Westport, CT: AVI Publishing, 1972).

155 *said one sales manager*: Conversation at SupplySide West 2011 conference.

156 *This is why the Clif Bar company, despite its best intentions*: Clif Bar confirms that it has been unable to buy suitable organic soy protein.

156 *Soybean crops produce at least twice as much protein*: "Soy Benefits," National Soybean Research Laboratory Web site, http://www.nsrl.uiuc.edu/soy_benefits.html.

156 *Whole soybeans are a tidy package*: U.S. Department of Agriculture National Nutrient Database, available at http://ndb.nal.usda.gov/ndb/foods/show/3147.

157 *fiber and vitamins have been processed out*: Solae confirms this nutrient loss.

157 *A serving of soy milk delivers less than 10 percent*: According to the ingredients pane on the Silk brand carton of soy milk.

158 *the FDA gave the go-ahead*: "Food Labeling: Health Claims; Soy Protein and Coronary Heart Disease; Final Rule," *Federal Register* 64 FR 57669 (October 26, 1999).

158 *The only real problem was that nobody could figure out why*: "Food Labeling: Health Claims; Soy Protein and Coronary Heart Disease; Final Rule," comment 37.

159 *Daniel Jones . . . wrote a letter to the FDA in February 2008*: Daniel Jones, President, American Heart Association, to the Division of Dockets Management, Food and Drug Administration, February 19, 2008.

159 *in 2004 Solae filed an FDA petition for an anticancer claim*: "Health Claim Petition: Soy Protein and the Reduced Risk of Certain Cancers," requested by Solae, LLC, February 11, 2004, available at http://www.fda.gov /ohrms/dockets/dockets/04q0151/04q-0151-qhc0001-01-vol1.pdf.

160 *William Helferich . . . fed groups of mice*: Clinton D. Allred et al., "Soy Processing Influences Growth of Estrogen-Dependent Breast Cancer Tumors," *Carcinogenesis* 25, no. 9 (2004): 1649–57.

165 *instead of Applegate's three heating steps*: Conversation with Tom Stone, director of marketing, Bell & Evans.

9. Why Chicken Needs Chicken Flavor

168 *a webinar she gave on global flavor trends*: Wild Flavors, "Emerging Global Flavor Trends and Innovations," webinar, June 15, 2012.

168 *ingredients in executive chef Jeremy Bearman's kitchen*: Conversation with Kristy Lambrou, Rouge Tomate culinary nutritionist.

170 *To complete the impression of grilled chicken*: This process for making "grilled" chicken is widely understood in the flavor industry and was confirmed by Tyson.

171 *she oversaw an edition of the artsy publication* Visionaire: "Taste," *Visionaire* 47, March 2006.

171 *Of the roughly five thousand additives*: Thomas G. Neltner et al., "Navigating the Food Additive Regulatory Program," *Comprehensive Reviews in Food Science and Food Safety* 10 (2011).

172 *Many ingredients that go into processed food*: R. J. Foster, "Flavor Mask-erade," *Food Product Design*, September 4, 2007.

173 *the $12 billion global flavoring industry*: This figure comes from John Leffingwell of Leffingwell & Associates (see the Web site at http://www.leffingwell.com), the best source for statistics on the flavor business. It's a sum of flavor sales ($7.4 billion), essential oils ($2.6 billion), and aroma chemicals ($1.4 billion). It does not include an amazing $27.5 billion worth of flavors created by the soft drink companies for use in their soda concentrates.

175   *boiled in large vats of hydrochloric acid for six hours*: The description of the standard process for making HVP and yeast extracts comes from Savoury Systems.

176   *It was cooked quickly, then frozen in a bag and eventually reheated*: Conversations with Dafne Diez de Medina, Innova's vice president of innovation, research, and development.

176   *When it began in Europe in the nineteenth century*: My account of the history and evolution of the flavor industry comes from several sources: Conversations with John Cassens, president, Cassens Consulting; Wayne Dorland and James Rogers, Jr., *The Fragrance and Flavor Industry* (Mendham, NJ: Wayne Dorland Publishing, 1977); *FEMA 100: A Century of Great Taste* (Flavor and Extract Manufacturers Association, 2009); "Our Heritage," Firmenich Web site, http://www.firmenich.com/m/company/about-us /history/history-1895–2008/index.lbl.

177   *In 1952,* Fortune *magazine declared:* "What Has Happened to Flavor," *Fortune,* April 1952.

178   *a biologist at the University of California*: Melinda Wenner, "Magnifying Taste: New Chemicals Trick the Brain into Eating Less," *Scientific American,* August 2008; Melanie Warner, "Food Companies Test Flavorings That Can Mimic Sugar, Salt or MSG," *New York Times,* April 6, 2005; Burkhard Bilger, "The Search for Sweet," *New Yorker,* May 22, 2006.

10. Healthy Processed Foods

182   *A 1945 photo in a book on the company's history*: Penford Products Co. 1894–2000: *A History of Progress* (Penford Corporation). All the other details about Penford come from conversations with people at the company.

184   *won't survive the heat, mechanical mixing, shear, hydrolyzation*: Conversations with Bryan Scherer, Penford's head of technology.

184   *The government recommends that adults get*: The Dietary Guidelines for Americans (Washington, DC: U.S. Department of Health and Human Services and U.S. Department of Agriculture, 2010), 41, available at http:// health.gov/dietaryguidelines/2010.asp.

185   *When resistant starches first started appearing*: Sources for this section on the politics of new fibers include Food and Nutrition Board, Institute of Medicine, *Dietary Reference Intakes, Proposed Definition of Dietary Fiber* (Washington, DC: National Academy Press, 2001); "Advance Notice of Proposed Rulemaking (Food Labeling) 72 FR 62149," Section F: IOM Report on the Definition of Fiber, *Federal Register,* November 2, 2007;

"AACC Report: The Definition of Dietary Fiber," *Cereal Foods World* 46 (March 2001); "Proposed Policy: Definition and Energy Value for Dietary Fibre," Bureau of Nutritional Sciences, Health Canada, December 2010.

189 *On a bright morning in March of 2010, Michelle Obama*: Eddie Gehman Kohan, "Transcript: Remarks By the First Lady to Grocery Manufacturers Conference," Obama Foodorama blog, March 16, 2010, available at obamafoodorama.blogspot.com/2010/03/first-lady-to-corporate-food-giants.html.

190 *the crowd stood up and flooded the room with applause*: Marian Burros, "Michelle Obama to Food Makers (Politely): Stop Fattening Our Children," Rodale.com, March 17, 2010, available at http://www.rodale.com/michelle-obama-speech-gma.

190 *On its Web site, the Grocery Manufacturers Association*: "GMA Statement Regarding HBO 'Weight of the Nation' Series," Grocery Manufacturers Association press release, May 11, 2012, available at http://www.gmaonline.org/news-events/newsroom/gma1/.

191 *a report done in 2011 by a former food-industry executive*: Hank Cardello, "Better for You Foods: It's Just Good for Business," Hudson Institute, October 2011.

192 *the Prevention Institute . . . looked at fifty-eight products*: Juliet Sims, "Claiming Health: Front-of-Package Labeling of Children's Food," Prevention Institute, January 2011.

194 *A store manager in Dallas reported*: Teresa Gubbins, "A Devil of a Time Finding SnackWell's," *Dallas Morning News,* October 20, 1993.

195 *Simple items like cheese, frozen vegetables*: E-mail correspondence with Hank Cardello, director of the Obesity Solutions Initiative at the Hudson Institute.

196 *Derek Yach . . . told public health lawyer and blogger Michele Simon*: Michele Simon, "How Junk Food Giant PepsiCo Is Buying Up High-Ranking Experts to Look Like a Leader in Health and Nutrition," Alternet.org, August 4, 2010, available at http://www.alternet.org/story/147738/how_junk_food_giant_pepsico_is_buying_up_high-ranking_experts_to_look_like_a_leader_in_health_and_nutrition.

196 *charged the Frito-Lay snack division*: Bruce Horovitz, "Frito-Lay to Make Snacks from Natural Ingredients," *USA Today,* December 28, 2010.

196 *Pepsi-Cola . . . had slipped to third most popular beverage product*: Charles Riley, "Diet Coke Fizzes Past Pepsi," CNN Money, March 18, 2011.

196 *Ann Gurkin . . . knew exactly why it had happened*: Duane Stanford, "Pepsi May Boost Marketing Budget to Take on Coca-Cola," *Business Week,* January 26, 2012.

197 *Pat Weinstein . . . put an even finer point on it*: Mike Esterl and Valerie Bauerlein, "Pepsico Wakes Up and Smells the Cola," *Wall Street Journal,* June 28, 2011.

197 *Nooyi quickly responded by redistributing*: "Pepsi Announces Strategic Investments to Drive Growth," Pepsico press release, February 9, 2009.

197 *Albert Carey told the* New York Post *about the new approach*: Josh Kosman, "Pepsi Big Bets on Company's Chips," *New York Post,* January 14, 2012.

198 *Mars's scientists quietly set out to understand*: Most of the details about Mars and flavanols come from conversations with people at Mars. Other sources include Jon Gertner, "Eat Chocolate, Live Longer?" *New York Times Magazine,* April 12, 2010; Alexei Barrionuevo, "Mars Expands 'Health' Line with Milk Chocolate Bars," *New York Times,* September 14, 2006.

11. Sit at Home and Chew

205 *Mrs. Ruffles, Ruby, and Treadmill Mom*: Laura Shapiro, *Something from the Oven: Reinventing Dinner in 1950s America* (New York: Viking, 2001).

206 *fix her family's meals in an hour and a half*: Ann Vileisis, *Kitchen Literacy: How We Lost Knowledge of Where Food Comes from and Why We Need to Get It Back* (Washington, DC: Island Press, 2007).

206 *barely thirty minutes of cooking a day*: "Society at a Glance—OECD Social Indicators," Organisation for Economic Co-operation and Development, 2011, available at http://www.oecd.org/dataoecd/38/52/47573400.pdf.

206 *dinner entrees made from scratch*: "America's Hurried Lifestyle Has Greatest Impact on Eating Behaviors over the Last 30 Years," NPD Group press release, October 18, 2010.

206 *frozen aisles have been supersized*: Ross Boettcher, "Frenzy at the Freezer," *Omaha World-Herald,* April 10, 2011, available at http://www.omaha.com/article/20110410/MONEY/704109890#-frenzy-at-the-freezer.

206 *Sales of frozen foods and beverages*: "Frozen Food in the U.S.," Packaged Facts, January 1, 2011.

206 *40 percent of women aged sixteen and over*: Calculated from "Quick Stats on Women Workers, 2010," U.S. Department of Labor, available at http://www.dol.gov/wb/factsheets/QS-womenwork2010.htm.

206 *studies show that women still end up doing the lion's share*: Joni Hersh and Leslie Stratton, "Housework, Wages and the Division of Housework Time for Employed Spouses," *American Economic Review* 84 (May 1994).

207 *increased by 74 percent since 1980*: "Statistical Abstract of the United States: 2012," Table 1337, U.S. Census Bureau.

207 *the group's president, Pamela Bailey, wrote*: Pamela Bailey, Letter to the Editor, *New York Times,* September 27, 2011.

207 *According to a Pew Research Center survey*: "Eating More, Enjoying Less," Pew Research Center, April 19, 2006.

214 *"I'm not one of those born healthy eaters"*: Just a Night Owl blog, at http://justanightowl.com/.

216 *after hearing about a workplace program*: The Full Yield, Inc., Web site at http://www.thefullyield.com/.

216 *national program teaching people how to cook*: All the details about Cooking Matters as a national organization come from conversations with people at the organization and from the Cooking Matters 2011 Annual Review.

Acknowledgments

Of all the people I am deeply indebted to for helping make this book a reality, my mom, Therese Warner, tops the list. Without her, none of this would have even been remotely possible. Not only did she plant the seeds of inspiration long ago with the food she raised my brother and I on, she aided in the book's creation over the past few years through unwavering support and hours upon hours of child care. Her insistence on eating some of my food experiments was troubling at times, but she approached it all with unfailing good humor.

My husband Rich was a constant source of love and encouragement. He taught me, a born skeptic, never to let go of the power of belief, and he tolerated my weekend disappearances with patience. Thanks to our amazing boys, Jude and Luke, who help me keep it all in perspective. One day they will understand, I hope, why Mommy went so crazy every time they tried to touch the strange food in her office.

I consider myself beyond lucky to have chanced upon an editor as gifted and diligent as Shannon Welch at Scribner. Her long hours matched mine, and the book profited immensely from her excellent editing. Nan Graham has been a beacon of enthusiasm, and I am fortunate to have her as a champion. My brilliant agent, Molly Friedrich, believed in the book from the start and has been a constant companion throughout the process.

I'm not sure this book would have ever gotten off the ground without Laura Rich's dogged insistence that we start a book proposal support group. She and Mari Brown refused to let me continue procrastinating on an idea that had been rumbling around in my head for years. I am grateful for their zeal and input.

I feel blessed to have so many smart friends who offered early edits on chapters. Thanks to Hannah Nordhaus, whose writing talents appear to be matched only by her editing abilities, Hillary Rosner, Florence Williams, Chris Hunt, and Tanja Pajevic. Sari Levy not only read chapters but also did countless hours of research. I was humbled by her belief in the project and benefited from her sharp insights. Christina Reilly also graciously contributed research and lent her video expertise. Lisa Ferri was a fountain of great ideas, and Chris Elam the bearer of a great Rolodex. I am grateful to Karen Leo, Ashwin Verma, Mariem Horchani, and Tim Warner for their love and support.

Thanks to the good people at Brewing Market and Vic's in south Boulder who let me spend so much time in their coffee shops that at times I felt as though I should be paying rent. I am indebted to their hospitality, cozy seating, and robust coffee.

This book seeks to follow humbly in the footsteps of some of the great works that have helped spark a more conscious and enlightened approach to eating. I have been encouraged and awed by Michael Pollan's *The Omnivore's Dilemma* and *In Defense of Food,* Eric Schlosser's *Fast Food Nation,* Felicity Lawrence's *Not on the Label,*

Barbara Kingsolver's *Animal, Vegetable, Miracle,* and Barry Esta-
brook's *Tomatoland.*

Among the people I interviewed but who do not appear in this
book is Scott Creevy of the Great Harvest bread store in Boulder.
Scott has been grinding wheat to make unbelievably delicious,
fresh, chemical-free bread in his store for twenty-five years. His and
other Great Harvests around the country serve as a reminder of
how letting technology take over doesn't necessarily mean a better
product. I am grateful to him for spending time with me.

A big thanks to Darcy and Cameron and the amazing Struckmeier family for letting me into their home. Their journey to
understanding what kale and tahini are is truly inspiring.

Finally, I am perhaps most grateful to the many food scientists
who spoke with me and graciously lent me their time. They will
undoubtedly disagree with many of the book's conclusions, but
I hope they will be able to see the importance of telling the story
this way.

Index